Decoding the World

About the Cover Art

The cover image and design, by Jarrod Taylor, is inspired by the Dutch surrealist graphic artist M. C. Escher, whose works *Bond of Union* and *Rind* were, in turn, inspired by H. G. Wells's novel *The Invisible Man.*

The genetic code inside the ribbon–the bases of As, Ts, Cs, and Gs–is not random. Here is the full code:

CTCTGTTAGCGTCTGCTCGTCAGCCTGTGAAGCCTG
CTCCTAGTACTGTAGACTCATATCCTA

Using a simplified grade-school DNA writer, this can be decoded into plain text:

NOTHING IS INEVITABLE

Decoding the World

A Roadmap for the Questioner

Po Bronson &
Arvind Gupta

The Convergence Trilogy, Book One

NEW YORK BOSTON

Copyright © 2020 by Po Bronson and Arvind Gupta

Cover design by Jarrod Taylor
Cover copyright © 2020 by Hachette Book Group, Inc.

Hachette Book Group supports the right to free expression and the value of copyright. The purpose of copyright is to encourage writers and artists to produce the creative works that enrich our culture.

The scanning, uploading, and distribution of this book without permission is a theft of the authors' intellectual property. If you would like permission to use material from the book (other than for review purposes), please contact permissions@hbgusa.com. Thank you for your support of the authors' rights.

Twelve
Hachette Book Group
1290 Avenue of the Americas, New York, NY 10104
twelvebooks.com
twitter.com/twelvebooks

First Edition: October 2020

Twelve is an imprint of Grand Central Publishing. The Twelve name and logo are trademarks of Hachette Book Group, Inc.

The publisher is not responsible for websites (or their content) that are not owned by the publisher.

The Hachette Speakers Bureau provides a wide range of authors for speaking events. To find out more, go to www.hachettespeakersbureau.com or call (866) 376-6591.

Print book interior design by Jarrod Taylor
Illustrations of authors' heads by Kristrun Hjartar

Library of Congress Cataloging-in-Publication Data

Names: Bronson, Po, 1964- author. | Gupta, Arvind, author.
Title: Decoding the world : a road map for the questioner / by Po Bronson and Arvind Gupta.
Description: First edition. | New York, NY : Twelve, [2020] | Summary: "In Decoding the World, Po Bronson and Arvind Gupta-two renegade venture capitalists from Silicon Valley-take everyday news headlines and decode them, leading us on a journey through their twisted and highly entertaining view of the world. Each chapter is prefaced with a real-world headline ripped from today's chaotic news cycle: Trump's trade war. Dying bees. Rogue planets. Beyond Meat. Glaciers melting. Bronson and Gupta then decipher what's really going on behind these headlines, and why. What they offer is first-hand experience in funding technologies to solve these problems, most of which involve genetic engineering. But what the authors then do with that premise is always surprising and unexpected. In one paragraph they are ripping it down to the bare bones physics or chemistry, and in the very next paragraph invoking history, philosophy, or psychology-while using literary devices borrowed from the surrealists, along with storylines from popular movies. The narrative holds a tightrope suspense, as we wonder what they'll do next, or what brazen thing they'll say. Decoding the World is the kind of book you get when you give two guys $40 million, a world full of messy big problems, a genetics laboratory to play in, and a set of Borges' collected works. After looking through their lens, you'll never see the world the same"– Provided by publisher.
Identifiers: LCCN 2020014841 | ISBN 9781538734315 (hardcover) | ISBN 9781538734322 (ebook)
Subjects: LCSH: Genetic engineering. | Genetic engineering–Moral and ethical aspects.
Classification: LCC TP248.6 .B76 2020 | DDC 660.6/5–dc23
LC record available at https://lccn.loc.gov/2020014841

ISBNs: 978-1-5387-3431-5 (hardcover), 978-1-5387-3432-2 (ebook),
978-1-5387-5419-1 (international)

Printed in the United States of America

LSC-W

10 9 8 7 6 5 4 3 2 1

Life will always be more work.
The only thing you can do is make it more fun.

Contents

About This Book

Decoding the World, at its core, tells the story of a friendship. It's a buddy story. It starts with Po coming to IndieBio–and ends two years later with Arvind leaving IndieBio.

Arvind created IndieBio as a laboratory for early biotech startups trying to solve major world problems. Glaciers melting. Dying bees. Infertility. Cancer. Ocean plastic. Pandemics.

There are some truths you can only learn through doing. And there are other truths you can only access by slowing down and really synthesizing. Arvind and Po embody this duality. Arvind is the fearless one, a radical experimentalist. Po is like the detective, patiently looking for clues others have missed. When they meet, their styles mix and create a quadratic speedup of creativity. Yin and yang crystallized.

The villain they're fighting is inertia. The status quo. They find scientists to create companies to solve important world problems. But it's hard. Capitalism fights back. One might say they are up against Isaac Newton's First Law of Motion, which says the bigger the mess, the easier it is to just keep going the same way we've always done it.

Unexpectedly, over the course of the two years, a classic Hollywood role reversal transforms them both. Arvind learns to think slower to build bigger. Po learns to act faster to see further.

As Arvind's departure draws near, he struggles to leave the sanctum he created. While Po has to prove he can keep the "indie" in IndieBio after Arvind is gone.

Throughout this book,  is Po, and  is Arvind.

All of the text messages are seen from the screen of Po's phone, with Arvind on the left and Po on the right.

Decoding the World

1

First Coronavirus Death in U.S. and New Cases Detected as Testing Expands Washington Post

Craig needed three days to start testing.

Akash wanted to start a clinical trial in 10 days.

Melanie needed 32 days to grow antibodies and sequence them.

Franco needed 45 days for his CRISPR test, which would drop the cost of testing to five dollars.

First went the handshakes. Second went travel. Everyone canceled their trips.

Even in a crisis, people want to look smart and rational. There was a compulsion across society to use the little we knew to declare predictions. It took several weeks to recognize the futility of looking into the future. The only honest people were those who admitted, *this is the unknown*.

There was no plan, just a way. We all went home from IndieBio, vacating the lab so Franco's team from Argentina could take it over and develop their test. After a day, we couldn't stand the feeling of retreat. This was not us. Our philosophy is action.

Then Arvind got the email from Akash.

Akash had a proposal to stop COVID-19. I forwarded it to the team.

Text from my sister, who is an eye surgeon at a major hospital in New York:

> Sad case today. A 17-year-old with a bad injury from yard work. Full corneal laceration, traumatic cataract and retinal detachment. His father was so devastated. When we were talking he collapsed and hugged me out of sorrow. Now I feel like I'm covered in corona.

Akash wrote that he had 735 kilograms of niclosamide on its way to the U.S. They were going to use it for a clinical trial as a form of birth control, as a spermicide. But they wanted to try it for COVID-19. They were in contact with the FDA. Niclosamide was invented by Bayer in the 1950s and was mostly used in the developing world to treat tapeworm. It had been pulled off the U.S. market in 1996, but it was still made elsewhere.

The argument was that niclosamide might work because, even though it had never been used to treat SARS or MERS during an outbreak, long after the fact it was shown that niclosamide could stop those viruses from replicating inside us. The COVID-19 virus had the same 11 proteins as SARS and MERS, the same arsenal of weapons.

Normally, VCs take weeks or months to study a deal. This took us about an hour to say yes, which was the length of our first phone call. It was a blind bet with no evidence, only theory. Usually those bets bust. We would have been fine with that.

We were worried about Craig. He'd been coming by the lab to borrow our RT-PCR for weeks. He was so frustrated by the lack of testing that he invented an alternative to the CDC's protocol. He took four steps and simplified it into one step. He and Gabe were out on the street, testing random people, including the homeless.

Then Craig got some positive samples from UCSF, and he was able to confirm his test worked. He had a lab space arranged. He could start testing in three days, but he needed money to buy a real-time ABI7500DX PCR to handle the volume. When we tried to wire him $250,000, he realized he hadn't even set up a bank account yet.

Sleep was important. Sleep as much as possible. Roll out of bed, Zoom calls for fourteen hours. Days became a photocopy of the day before. Without the texture of moving around the city, or bouncing around IndieBio, I became forgetful; my memories were unmoored from geography, synthetic, like boring dreams I couldn't wake from.

Arvind wrote a letter to all our alumni:

> "Nobody knows how long this will last. Nobody knows how this will end. I do know this. Nothing is ever as good as it looks or as bad as it feels. We will be fine. We may even be better from it. But first, we must survive it. You are all IndieBio companies. Which means you are already survivors. Born and bred in the basement on Jessie Street in the Tenderloin. Walking into the office meant stepping over the hardships of life just to face it again in the lab. But this shock is different. Covid-19 will test us all."

At least a dozen of our alumni were running into the fire.

In Korea, their Zoonotic Virus Lab tried 3,000 existing drugs, in a high-throughput screen using kidney cells infected with the virus. They found 24 that worked well, and niclosamide was one of the two that stood out above the rest. We called the Gates Foundation to get it on their radar. Around the same time, Akash got great results back on niclosamide from the Galveston National Lab, a Biosafety Level 4 containment facility that worked with live virus. Three medical centers had agreed to start a trial.

Text from my sister in New York:

> I have an emergency transplant to do. She will lose her eye by tomorrow. Patient should be tested for COVID but hasn't. The virus becomes airborne during intubation and can infect the whole operating room for hours. They don't have a mask for me.

I sent an N95 mask to her by FedEx. It arrived three hours late.

We had just opened IndieBio New York. Our team there never even had a chance to visit the lab before being sent home. But they were working with the state, and it became clear we needed to send Craig to the New York Genome Center in SoHo. His ability to do high-volume testing was more needed there than in San Francisco. He started with all 4,000 people at the United Nations.

Working with Sean O'Sullivan, the managing general partner of our firm, we decided to announce publicly that we would fund eight COVID-19 initiatives.

We got drowned in applications. Everybody needed money. Few VCs were open for business. The team did fifteen to twenty Zoom calls a day. Decoy strategies at the heparan sulfate receptor. Protein degraders. Vaccine platforms. Llama antibodies. Antivirals from plants. Sanitizer tech that used nothing but water and charged ions. I've never seen a team absorb so much information, so fast, and make decisions on the fly about what they believed was our best chance.

Exosomes, engineered to rescue lung function in patients who can't breathe. Mickey needs 60 days. We're in.

Some people checked the stock market ten times a day. I checked Nextstrain to follow the mutations and migrations. Both are a kind of Rorschach test.

Message posted on IndieBio Slack:

> Also if you join by video, please dress up. Haha. Theme is that it is the year 3020 and the human race has been in isolation for 1000 years.

Message posted on IndieBio Slack:

> Shit! Ryan Gosling's Butter Sacrifice party is popping! The tunes are crazy!

Franco called from IndieBio. Half his team just tested positive.

Everyone says this started in bats. And that as long as there are bats, there will be viruses spilling over into humans. But parts of that story are missing.

Bats are the only mammal that flies. Flying raises their heart rate so high (up to 1,000 beats per minute), and they burn so much energy, that the DNA damage created would kill anything else. So along with the ability to fly, bats evolved more powerful DNA repair mechanisms. Their genome has a second copy of P53, the guardian of the genome, which patrols their DNA for mutations. They also express far more interferons. These protect and repair the bat genome so well that bats can *also* handle all the genetic chaos that viruses create. The viruses will live in bats, without tearing them apart.

Once these viruses spill over to humans, it's like LeBron James showing up at the local pickup court. It's too easy. The virus has evolved to compete against far superior bat defenses. Against our weaker defenses, the virus carves us up.

But what's missing is this: Bats don't normally infect humans. That's why this happens rarely. Bat genomes are so good at keeping viruses in check, that most of the time, bats are no danger to humans.

It's only when bats suffer immunological stress, and their viral load goes way up, that they get sick and can pass their viruses to humans. One of the most common stresses, recently, is loss of habitat. Deforestation, urban development, and arid wetlands.

So this didn't start with bats. This started with whatever caused the bats to get sick.

Everyone is looking for a drug. But the body is better and faster at designing drugs than any pharma company.

Mutation is normally bad for the body. But we actually have little tiny laboratories in our bodies where we turn on hypermutation when something foreign gets in, like a virus.

We do this in a controlled, safe setting–on a particular stretch of the genome in B lymphocyte cells. This is where the body invents antibodies. The hypermutation is called V(D)J recombination. Variable, Diversity, and Joining.

Every time the B cell divides, short code chunks of Vs, Ds, and Js randomly recombine. The genetic proofreaders don't interfere. The body keeps recombining and recombining until–randomly–one works. If an antibody latches on to the virus, all sorts of signals ramp up, and cells clone the antibody rapidly, like a drug factory. These antibodies mark cells for destruction.

Every single person who gets the virus has to invent their own antibodies. It's a race: Can your body hypermutate a drug to save you before the virus turns your blood vessels into pink slime?

Someone started tailing me. I tried to shake them in the baking aisle. At the grocery store, it was like a zombie film. In the vegan aisle, the shelves were full. I lost him there.

In Iceland, my family was losing their jobs. The government was testing people randomly, calling them out of the phone book. Ten percent of the country had been tested, so the media was reporting how Iceland hasn't sheltered in place. But according to our family, nobody was leaving the house. Except to go out and shoot caribou.

I missed being at IndieBio. And even though I wasn't going to actually leave IndieBio for three more months, being sheltered at home made me aware of what I'd miss the most. It's how the day starts–how almost every day at IndieBio starts.

I park across the street, step over a whiskey bottle and a syringe, give a hug to whoever opens the door, and descend the steel staircase to the Ivory Basement. The distance to my desk is about the same as a fashion show runway. High fives and more hugs, quick updates, then the team rolls out for coffee. We sit outside on Market Street, the city's raw spectacle rolling in ecstasy at our feet. We tell personal stories. The drama at group houses. The parties we can't unsee, even if we wish to. We triage the companies we're incubating. It feels so good to spend time together. Then we start brainstorming. Maybe it's how plant cells use gravity to know up from down. Or tax policy in India. Or someone declares the amount of joules in the chemical bonds of a pound of body fat. Somehow, the conversation leaps topic to topic with every exchange, and we always end up talking about something we never could have predicted even a minute earlier. It doesn't get better.

Melanie had made antibodies against Zika when she was at IndieBio. She believed she could do it again against this virus. There were 23 companies trying to invent the perfect antibody, and a dozen more trying to design one with a supercomputer. "They're slow," Melanie said. "They start in humanized mice, then hope what they get works in humans. It takes tons of repetition and adjustment. In a computer, you can design an antibody in days. But you have no idea if it will have off-target effects, so that testing takes months and months." Melanie said she could do it in 32 days.

Melanie's approach was truly unique. Her company, Prellis, was the world leader in 3D-printing human organ tissue, using lasers and stem cells. Her goal for Prellis was right out of a science-fiction movie: She wants to print a new liver for patients when their liver is shot. She was getting close. She'd been making mini-livers.

But this was wartime. "I can make dozens of mini-lymph nodes, little immune systems. I'll inoculate them with the virus. They'll create antibodies just like they would inside a human body. I'll screen the antibodies for which works best."

In a week, the lymph nodes were printed. A week later, inoculated. One more week, and the miracle of V(D)J recombination was generating antibodies to the virus.

Collectively, the search for remedies is like trying to defuse a bomb that you're carrying in your own hands–while simultaneously running a marathon.

Franco says his team is *not* positive. It was a false alarm. All the more need for his far more accurate test.

My daughter asks me how talking on the phone can help stop the virus.

One day we will go back to normal. But I'll fondly remember the girls Zoom-bombing me on my calls, taking family walks in the rain, and losing "name the animal" to my five-year-old.

2

"World's Supermarket" Returns as Epidemic Eases in China The Star

Once upon a time, I lived and worked in Shanghai, China.

I saw a country, an entire nation, in the midst of rebuilding itself, almost overnight. It was a society coming out of poverty.

It was staggering how much was possible with so little. The sheer scale and speed of this reshaping was warping reality.

My wife and I lived in a local housing development in a Chinese neighborhood. We had one room in a town house, sharing it with two brothers and an old lady. When the temperature dropped below freezing, she would fill the kitchen with steam from a pot, her nose bright red. Thirty tightly packed town houses were divided by narrow lanes that converged on a single exit gate, where a twentysomething guard lived with his family in the gatebooth–a four-foot-by-six-foot shack. It was made of peeling plywood; they had a single hot plate and a bed that folded down from the wall and was held by a chain. In the winter it would freeze, rain, and snow. Every day, the guard's family would greet us with joy, as we biked out onto Wuyuan Lu and past the wet market on the corner.

I was a designer. I was designing a new beautiful, curved smartphone for Samsung that would cost more than most people I knew in China could afford. I struggled with that. I knew something was being missed. As we crafted the high-end, high-margin blockbuster products for the wealthy, another world was taking shape.

Three hours south of Shanghai was Yiwu. Yiwu market was a mall of wholesale trade goods beyond the imagination. Everything was sold in bulk, in little stores, with factories behind the stores. We walked ninety minutes straight across the mall and never saw the other side. Yiwu had 75,000 stores–150 *times* more than the biggest malls in America. It's called "The World's Supermarket." A cameraman from CNN spent four days there, and he couldn't traverse it all. Knockoffs of everything you had ever seen, and everything you had ever imagined, were for sale in bulk, in ten-thousand-unit lots.

You may never have seen Yiwu, but if you've traveled at all, you've seen what they make. All over the world, in markets from Paris to Jerusalem to Los Angeles, are products made in Yiwu. Later, in Istanbul's Grand Bazaar, I saw Turkish coffee cezves for sale that I had seen previously in Yiwu. And in Kolkata, India, I saw merchants selling sandals that I had also seen in Yiwu stalls. Yiwu is the backbone of the world economy.

How exactly it was the case that 90 percent of everything in the world came to be made in China, I wasn't sure. In 1982, the government laid cement boards over a ditch in Yiwu so the first stalls could be erected. We blinked, took a nap, and somehow, it turned into this, a scene from a sci-fi novel on another planet. A new Silk Road had been paved in cheap disposable goods. "Cheap" was the operative word. Most of the world didn't have much money. They loved cheap.

"Shanzhai" was their word for it. To copy. Counterfeit, without shame. It was not just to replicate products. It was to replicate a notion of living, influenced heavily by the West. I saw entire suburban tract mansion suburbs pop up in months. McMansions. Children on bicycles. Everything but the milkman and Girl Scouts at the corner selling lemonade.

I remember coming back from Ningbo, across Jiaozhou Bay. Halfway across the world's longest bridge, out of the fog rose a tower known as Haitian Yizhou, "the land between the sea and the sky." The tower was shaped like an eagle. To design such a bridge, to conquer the strongest tidal forces on the planet, took engineers ten years to figure out–before construction began. Why they added to the complexity by erecting a monument that reached into the sky confused me. That was the moment my friend taught me about Da Qi.

Da Qi is a concept that is wrapped up in Confucian masculinity. It's a kind of generosity of spirit that comes from economic success. It's almost two ideas at once–both financial prosperity *and* sharing its bounty. Da Qi means leaving your village and coming back rich. Paying for dinner is a gesture of Da Qi. Giving expensive gifts is a gesture of Da Qi. The government showed its Da Qi everywhere it could.

The Chinese loved massiveness, but they understood that massiveness in a building creates no awe without empty space around it. Da Qi is measured in empty space. The Palace Museum in Tiananmen Square is surrounded by 109 empty acres. The pagoda at the Temple of Heaven is surrounded by 640 acres–six times the size of the National Mall in Washington, DC. Space, empty space, created authority, impossibly grand.

Even in a corner convenience store, merchants would make piles of goods by the entrance, to convey a sense of incredible abundance.

The rebuilding of China was happening before my eyes, in my three years there. Almost none of the buildings and bridges and roads that I used every day had existed a decade prior. It made me believe that transformation on a large scale was possible. A billion people at once could just start over, do it another way.

It was not the China we hear about today, where the government has an eye on everything and everyone. Back then, eight years ago, this world was being built ad hoc, organically. People were risking their livelihoods and their life savings to take the risk to finance it. Yiwu didn't arise out of a master plan. It grew by experiment. Each stall was an experiment, until the stalls became stores, and factories. It grew fast, it grew blindly, and it grew dense.

China was evolving so fast not because of its master plan, but despite it. A billion people amounted to a billion experiments.

I had a religious studies tutor. He was a monk and a scholar. We met once a week in cafés, and sometimes he came to my office. I was interested in the helix of interaction between Confucianism and Buddhism during the first century CE along the original Silk Road between Northern India and China. It was during that time that Confucianism, which was historically very pragmatic, absorbed some mysticism.

But I was struggling with the clash between my spiritual lessons and the dynamic experiment I saw outside on the streets. The fundamental precept at the core of both religions is that the *mind* is the path to knowledge and improvement. That wisdom will emerge if you sit still and discern.

There wasn't a lot of sitting still outside; there was a lot of action. I got frustrated at the chanting and recitation the monk made me do. He didn't want to be argued with but I did anyway. He didn't like me questioning him. Finally, perhaps to appease me–to find a bridge between my restlessness and his sacred texts–he laid a quote on me, from a Confucian scholar around 300 CE named Xunzi.

"Practicing is greater than knowing." Not *better* than knowing. *Greater.*

The philosophy of those words may not be self-apparent, so let me unpack them. "Practicing" means *trying*, it means *acting*, it means *doing*. Xunzi–and many philosophers since–made the case that the *Doer* learns things beyond the reach of the *Knower.*

It made inherent sense to me, crystallizing a lot of what I'd been riddling on, including the tension between master plans (made by a Knower) and bottom-up self-organization (all the Doers). In me emerged a philosophy of action. When you act, you learn. That kind of wisdom beats the knowledge you read in a book, every time. To really seek answers, you need to act. To really develop your mind, run more experiments.

The night before we left, the old lady downstairs made us dinner. We called her "Teacher Wang." She took us by hand along Wulumuqi Lu to the wet market, which was under the cover of corrugated steel panels. Everything for sale was still alive, except for the pig faces, peeled off and hung to dry, like masks. Writhing eels, huge toads, coiled snakes. Teacher Wang chose a three-foot-long ribbon fish, which the fishmonger gutted and cleaned, rinsing his cutting board onto the concrete. We tried to pay, but Teacher Wang refused. Minutes later, the fish was chunked and dusted with rice flour, sizzling in Teacher Wang's wok. Then she covered the fish in a simple braise, with garlic, ginger and spring onion, dark vinegar, white pepper, and soy sauce. She talked about being a teacher, and about her son, and the fortune of the spring weather.

In the morning, when the truck was loaded with our small room's worth of belongings, the old lady downstairs cried and hugged my wife. We stopped at the guard's shack. His girls dropped their piecemeal needlepoints to rush outside.

3

Silicon Valley's New Obsession: Boring-Ass Startups

The Hustle

IndieBio erupted just as the Bay Area was kind of getting sick of itself, whining about how everyone was on the make and everybody's startups were lame. People who worked at Facebook would go out to dinner and express regret that their job was really to sell ads. Everyone made themselves feel better by hosting Social Purpose Parties. This is where you drink, give money, and talk about using blockchain to protect the environment or end poverty. There was a hunger in the Valley for something rad, something uncompromising. Saving the world was everyone's favorite topic; they talked about it endlessly. But talk grew cheap.

IndieBio wasn't a Think Tank. It was a Do Tank. That's what was so refreshing about it.

What Arvind was doing at IndieBio those first years in San Francisco was felt by all but articulated accurately by none. Yes, *technically* he had created a venture capital fund. But it was more like he had stolen the tricks of capitalism, or misappropriated them, and was using them to rebel against capitalism.

IndieBio had the kind of public profile in San Francisco that a hip-hop record label might have in LA, spinning out emphatic, provocative hits. I'm pretty sure IndieBio was the only venture capital firm with *fans*. Fans who rooted for their companies to succeed. At night, hundreds of people would show up.

Nobody was making money yet, but they were lighting up the public imagination. At one of the very first public IndieBio events, Arvind was quoted by the *Guardian* of London saying, "I'm one of the only people in the world that's eaten a dinosaur." Technically, it was a mastodon–and it wasn't mastodon meat, it was a mastodon gummy bear. Which a startup had made from ancient mastodon DNA to show off that animals don't need to be slaughtered to have gelatin for gummy bears. But all over England, people were reading about a guy in San Francisco who eats dinosaurs.

And then IndieBio created the first lab meat company. Hamburger in a petri dish. As a society, we're getting slowly accustomed to this concept, but when it first came out, five years ago, it was so provocative. We looked to the wonders of biotech when our lives were in danger. Not when we were hungry. The idea of eating something made with biotechnology was totally foreign and weird.

The companies kept coming, wild and outrageous, defying expectation. If you can make meat without the cow, then you can save the forests by making wood without the trees. And if you can do that, then surely you can make plastics that will never pollute the ocean. We can stop overfishing. We can make leather from mushrooms rather than animals. We can stop throwing away our food waste and turn it into renewable power. We can save the bees. In the hands of the scientists who came to IndieBio, it felt like anything bad could be magically transformed into anything good.

The message was heard by those who were hungering for it. *Hey, we're rebuilding the world. Join us.* One cell at a time, one organism at a time. All life-forms were up for grabs. Not rebuilding it *virtually*, not making a digital version–rebuilding the real world all around us.

Hey, we're rebuilding the world. Come join us.

If you ever went to IndieBio, you were in for a surprise. Because when you heard "San Francisco Venture Capital Firm," you imagined some huge swank office on Rincon Hill looking out over the Bay Bridge. And even when you imagined "Biotechnology Lab," the mind conjured the carefully manicured, perfectly architected UC San Francisco biotech center.

IndieBio wasn't an ivory tower. It was an ivory basement.

IndieBio was on Jessie Street. Jessie Street is basically around the corner from Crack & Whore, where everyday street life included dumpster diving, gutter syncope, and sidewalk vomit. IndieBio is below a methadone center. It's next to an apparel sweatshop factory. Across the alley is a massive steam boiler that serves all downtown with heat.

When you finally find the IndieBio door, getting off Jessie Street is a relief, and for a moment it's a sanctuary. The main floor is a nice loft space, with thirty-foot beams, spiral staircases, and metal catwalks. It was probably a dance club in a previous life. But then you learn, this floor *isn't* IndieBio. IndieBio is downstairs, in the basement, where the ceilings are low, and the desks are jammed close together, and the lab glows brightly.

And you slowly realize that this whole setup for IndieBio was *intentional.* Arvind built it this way. As a fence, an intimidation forcefield. To scare off all the pretenders. To make sure only the hardcore and dedicated came. To make sure it was incorruptible. IndieBio wasn't there to make anyone feel comfortable. Arvind sat in the back, in the last row of the cave. He had no office.

Strangely, corporations and governments braved this intimidation forcefield regularly. They'd pull up in their Mercedes Sprinter vans and unload a gaggle of top executives, often speaking foreign languages. All around the world, economies were sagging with the burden of unsustainability. Nations were worried how they were going to feed themselves, or how they were going to compete with China's scientific power. And so they came to IndieBio to learn, to look, perhaps to be inspired. IndieBio became one of the favorite stops on the corporate tourism itineraries. Pretty soon Bill Gates had invested, and Jeff Bezos had invested. Celebrities came for special, private visits–but were treated no differently.

At first IndieBio was known for saving-the-planet stuff. But then they started moving into other industries, including ones close to the Valley's pockets. One company got onstage and basically told the computer industry that they'd been doing it wrong for decades. Computers were a lousy way to store data–you could store it a million times more efficiently if you used DNA. Another showed how computer chips would be better if brain neurons were wired in–they flew a drone using dragonfly neurons–and this was before almost anyone had heard of brain-machine interfaces. Silicon Valley started sending rockets up to the International Space Station, so IndieBio sent worms aboard. Worms that would practice terraforming Martian soil. Worms that some IndieBio scientists had figured out how to communicate with, using pheromones. Worms speak pheromone.

One entrepreneur got up onstage and said, "The future isn't organic. It's synthetic." Sometimes, you didn't know if these companies weren't pranks or performance art. To stop rhinoceros poaching in Africa, IndieBio had a company that was going to flood the market with synthetic, 3D-printed rhino horns. Made with actual lab-grown rhino keratin. Exact in every way, just that these rhino horns had never been attached to a rhino. Maybe they should have tried mastodon.

Before Arvind started IndieBio, he went to present his idea to the most successful venture capitalist of all time, Mike Moritz at Sequoia Capital. Mike was so successful that he'd been knighted by the Queen of England. Moritz was nice to Arvind and heard him out. Arvind was desperate to know what Moritz really thought, so afterward he pressed a friend who knew Moritz for more intel.

"You don't want to know."

"I do," Arvind begged.

"He said, 'Your friend is on a suicide mission.'"

San Francisco had always had biotechnology. Genetic engineering had been invented here. But there had always been a cozy relationship between professors at the top medical centers and venture capital. When a big *Nature* paper came out, the mandarins of venture capital took the technology and gave the professors a tidy number of shares in a company they created.

Some biotech VCs even had offices in medical centers. All the great deals in biotech were spoken for *before they even existed*. They weren't going to let a newcomer like Arvind in on the bounty.

But Arvind wasn't betting on the professors. He was betting on the graduate students. All across the country, university labs were overflowing with scientific talent. They might love to be professors, but there weren't enough professorships to go around. The odds of becoming a professor–even for a highly regarded postdoc with eminent journal papers–was only one in six. That meant five out of six needed somewhere to go. Jobs in pharma paid well but weren't exciting. These PhDs often had ideas for startups, but few venture capitalists were willing to give these young scientists the money to start a company. That bias was Arvind's opportunity. Arvind funded them in droves, and he knew how to build companies. Somehow he knew what it actually took to get shit done. And when one scientist came to IndieBio, she would email all her friends back in academia that they should come, too. By casting this wider net–looking for talent where many others didn't–IndieBio had become one of the most diverse places in Silicon Valley. This, too, was yet another way that visitors to the ivory basement felt like they were standing face to face with the future.

Traditional pharmaceutical biotech was big band music compared to IndieBio's punk swagger.

If you measured the amount of money running through IndieBio, it was a pimple compared to traditional biotech. IndieBio was investing only about $10 million a year. But if you counted the number of companies, it was a different story.

Each year, Arvind was launching twenty-five new companies–fifteen more than any other traditional biotech venture firm. By the fourth year, IndieBio had created 105 companies. In that time, IndieBio had created about as many biotech companies as the graduates of all the University of California campuses combined had created; and double the number from Harvard, Stanford, or MIT.

Which is a way of measuring this universe that a *lot* of people would scoff at, if they wanted to feel superior. In the ecosystem IndieBio survives in, having lots of companies isn't as important as having valuable companies. Some people looked at IndieBio the way they'd look at the kids at a pitch contest, or a startup competition, where the winner takes home $5,000 from the sponsor. *Hey, that's cute.* And Arvind knew it. It ate at him. It drove him. *Just because our companies are small doesn't mean they won't grow up to be as big as you.* He never said that, but I could feel it, from the day I arrived. In that first year, our companies grew in value from $600 million to $1.4 billion–a kind of unicorn salad, if you will. Once they were over a billion, nobody could scoff anymore. A billion dollars of value is when Silicon Valley takes you seriously.

I didn't go to IndieBio to save the world. I went because they were supremely cool. What I saw in IndieBio was a movement. It would inevitably be a movement that, like all social movements, soon leaves its origins behind. IndieBio would be cloned, or replicated and improved upon by others. But IndieBio had created the genre. IndieBio was the original.

By the time I joined IndieBio two years ago, I thought I was late, that maybe all the cool stuff had already happened. Like someone who only started listening to Modest Mouse when "Float On" hit the radio waves.

But the suicide mission wasn't over yet. During his twenties, for thrills, Arvind was a BASE jumper. He once jumped off a New York skyscraper. Supposedly, he had long given it up before he went to China.

But I would come to realize he had *never* stopped BASE jumping. He was still doing it. He was just taking us all at IndieBio with him.

4

33.48 Tons of Dead Fish Collected in Pinellas County as Red Tide Bloom Lingers ABC Action News

It was almost biblical. It wasn't just fish. Thousands of dolphins, manatees, and sea turtles also died. Runoff from the sugar plantations, combined with shifting current flows from the Florida watersheds, triggered a massive bloom of red algae that lasted eight months. The algae took all the oxygen out of the water, suffocating marine life. A friend of mine came across a wave of dead tarpon, each a couple hundred pounds, floating like logs. The stench was unbearable. As the algae hit the shore, the waves broke the cells open, releasing neurotoxins into the sea breeze. Many people showed up in hospitals with burning sensations in their skin, unable to breathe, suffering vertigo. Fishermen were paid to haul up dead fish before they washed ashore.

All over the world, some version of this is happening. As we disturb the planet, microbes get to work, feeding on pollution, sucking up our oxygen, or chewing on melting permafrost biomass and releasing their methane. Rising sea levels bring more than just water ashore–it's like opening the gates at the germ zoo. It's not the sea level or the heat that will get us; it's the microbes and viruses.

There's a new superfungus in our hospitals; it kills half the people it comes in contact with. It used to be harmless to humans because it couldn't survive in our 37°C bodies; we were too warm for it. But thanks to climate change, the superfungus was coaxed along, evolving to adapt to higher temperatures, and then infecting people–first in Japan, Pakistan, and India, then South Africa and Venezuela, before appearing now in the warm confines of the United States.

I've read a million articles about climate change, and not a single one explains the most basic point: Temperature, in living systems, is crazy critical to the chain reactions of biochemistry. In our lab, we have about three hundred pieces of equipment. At least one hundred of them control the temperature. Temperature *matters*. There's a reason we work at 98.6. It's called "molecular physics." Truly every living thing evolved to work at a certain temperature. You change the world's temperature, you stress all living things, from the quantum level up. Plants and animals evolve slowly. Microbes evolve in days.

The fundamental, existential question of our times is why society can't agree that we have to radically address climate change.

Solutions exist–but we're not committed yet to using them. So the most important question of our times is not "How are we going to save the planet?" and is instead, "Why don't we all agree to do so?"

What truly keeps us from such an agreement?

Everybody I know points their finger at climate deniers. And this pisses me off.

It pisses me off because if we were *just* fighting the climate deniers, we would easily win.

What we don't understand is our own culpability in filling the world with confusion, smokescreens, and illusions. Much like the boy who cried wolf, when we sound the alarm on climate change, we don't realize that it comes across—to the rest of society—no different from any of our other wild, science-y predictions of the future. About which society is both entranced and skeptical. Over time, by constantly making disorienting predictions about the future, we have trained society to handle these predictions the only way they know how: to wait and see.

Futurism has become a dangerous game. In the '90s, I was at *Wired* magazine. They predicted the future would have cryptocurrency, virtual reality, video chat, smart watches, e-commerce, artificial intelligence, robots, movie streaming, Mars rovers, solar power, carbon recycling, social networks, corporate space rockets, and a drastic reduction in infant mortality worldwide. All of that happened. Some of it took a while. A lot of other things did not come true, and a lot of people lost money betting on the specifics, but on the whole an insane amount of wealth was created—so much money that nobody was rich enough to ignore it. Now, everybody is listening.

Think for a moment about what people *not* in the Silicon Valley ecosystem are hearing out of Silicon Valley today:

- That we're going to live on Mars.
- That artificial intelligence will turn evil and wage war with humans.
- That the oceans will rise, deserts will expand, and the food system is in jeopardy.
- That the robots will take tens of millions of jobs.
- That we'll print replacement organs for transplants.
- That we'll eat beef meatballs without killing cows.
- That rich people won't age.
- Everyone's going to get $1,000 a month, for nothing.
- CRISPR will lead to a new transhumanist species.

Taken together, it's a pretty strange future being painted. And the natural reaction—probably the safest option—is to wait and see.

In the context of all these proclamations, climate change has a hard time standing out as different—as not just another confusing possibility that triggers healthy skepticism.

It's not intentional; it's an accidental side effect. It comes from good people doing good work trying to solve massive world problems. But when those good people cry out for climate policies–and I'm going to include Arvind and myself in that–we may not realize we are partly responsible for the confusion. We have meatball companies. We have organ-printing companies. We have longevity companies. We have robot companies. We do crypto. We have space-agriculture companies. We do all that. We have great friends who are transhumanist advocates.

Silicon Valley actually operates a lot like Donald Trump. It says a heck of a lot of things, and the rest of the world can't tell if it's bluffing or real. Is it just hype? Is it serious? Is it wishful thinking?

Plenty of warnings about pandemic risk had been issued, but little was done. Microbes finally got their 9/11 moment. The world will never be the same. Our financial system had its 9/11 moment in 2008. It was sorely needed, and it triggered some pretty serious reforms. Now the climate needs its 9/11 moment.

Without such a disaster, what happens is we get habituated to ever-escalating phenomena and just incorporate it into the new normal.

The bar for an awakening moment is pretty high. It's remarkable how much has already been tolerated without triggering a collective mobilization. Anthrax being released by the permafrost? Oh well. A sudden rise in flesh-eating bacteria populations on the coastlines of the American South? Oh well. Lake Erie turning bright fluorescent green with algal blooms? Oh well. Houses in Greenland collapsing into the earth? Bummer. The Ogallala Aquifer that our breadbasket states rely upon for water is drying up? Just add it to the list. Kansai International Airport in Japan is underwater? Meh. A piece of Antarctica the size of France is about to go? Text me when it does. There's a new virus in China? That's China's problem. There's thirty-three tons of dead fish washing ashore in Clearwater, Florida? Maybe that place needs a new name.

It's all the new normal. Wait and see.

Instead, the industrial-political complex sees an opportunity. Ninety percent of the world's population, and almost 70 percent of the land, is in the northern hemisphere. Without an Arctic in the way, shipping that $18 trillion in global trade is getting a lot easier and faster. Crossing the oceans is no longer necessary—just take the polar route. Cut weeks off the trip.

I'm not sure what would trigger an awakening moment on climate. New Orleans is only one foot above sea level (on average), and yet also the second-biggest oil-and-gas industrial center in the United States. I have a lot of family there, and they lost seven out of nine houses in Katrina. The levees are taller now. The human capacity for cognitive dissonance is infinite. Especially when it happens gradually, and even when it happens all of a sudden. They're keeping their eye out for hurricanes on the Gulf, when the real hurricane is coming from the high-rises on Poydras Street, where the oil and gas companies have their boardrooms.

Nobody thinks the world is without problems. The problem is we can't agree on the problem.

Meanwhile, Silicon Valley is out here doing its usual thing, trying to solve everything at once, but accidentally spreading mass confusion and plenty of sparkling distractions. Silicon Valley purposefully and intentionally creates a bubble. It's an ideological bubble, where it's safe to experiment on the improbable. Where mere illusions have the chance to become reality. Only by taking on the impossible can we learn what's possible.

This realm of illusion is spread by our creations. The secret appeal of social media was its meritocracy. There are no admissions criteria. To play, you don't have to be from the right school, or have grinded away at homework, or bribed the right people. That right there is a huge antidote, a relief, compared to the real world. Everyone gets to try their turn at the game of accumulating followers. All you have to do is get attention. That's the one and only rule. Players can present themselves any way they want, as they want to be seen. Reality is tenuous; it's merely a dimension that players can opt out of. Filters, lenses, fakes. Identity becomes a key question, and a movable feast. Who are you, when you can be anybody? Soon, the "real me" exists there, not here.

So I'm going to argue an unusual point here. Which is that Silicon Valley, in toto, has become a player in its own attention game. Silicon Valley is addicted to attention just like some two-bit Instagram celebrity. Just like a striving influencer, Silicon Valley desperately wants to be liked. It wants followers. It wants clicks.

But there are times where Silicon Valley isn't doing anything really worth talking about. It's making loads of money but on boring-ass enterprise software. So this tension leads Silicon Valley to start talking a lot about cool future technology *as if it's going to be a big deal very soon*. When in fact it's not even close to being ready and is more than a decade away, or several decades. But they can't help it, because they're desperate to be the center of attention again.

You know how every car side mirror is required to state, in fine print, "Objects in Mirror Are Closer Than They Appear"? Well, Silicon Valley should carry a warning sign with every pronouncement: "Innovations Discussed Are Farther Away Than Made to Appear."

My term for this dynamic is "Eating the Future." It's like a village that breaks into its winter food reserves when it's only October. All these innovations from the far-off future are being "consumed" and "digested" in the media, at great length, long before they're remotely real.

People are freaked out. Our world is increasingly in a kind of incomprehensible code. Our world is mutating, in every dimension, from the physical world of our planet, to the social fabric we rely upon, to the very fundamentals of humanity we grew up taking for granted: country, family, work, gender, health, death.

It's against this backdrop that this book arrives, from two Silicon Valley insiders. And this should be the part of the book where I explain why we're different, and why you should trust us, or maybe why we're more reliable than everyone else.

I tried writing that argument, but it felt false. Because the truth is, it doesn't matter who we are. You *shouldn't* trust us. Not yet. You should absolutely be in "wait and see" mode.

Any confidence you have in us, as writers/thinkers, should be earned. One chapter at a time. We thoroughly encourage you, at this early stage of the book, to be on high bullshit alert. Be a skeptic–it's smart.

The honest truth is, even we don't know if we're capable of writing this book.

Before I came to IndieBio, I had spent four years as a futurist consulting major corporations on how the world was changing. The more I stared into the future, the more trouble I saw ahead. I was desperately seeking a roadmap to navigate our way through. At IndieBio, I quickly learned there was no such master plan–but more than that, it was wrong to think this period we face could be planned out at all. It was fast, blind, and dense. Nobody is an expert in the unknown. Planning wasn't the way to solve it. Experimenting was.

There is no plan, just a way.

We'll take wait and see.

●●○○ 10:41 PM 100%

< Messages **Arvind** Details

On the television, the future is always creepy. Black Mirror, Twilight Zone, etc.

I think what makes the shows creepy is that every episode, there are new characters. What happened to those people in the last episode? They're gone.

Hahah.

What's creepy to me about the shows is they always present the humans as sheep.

Passive and accepting.

Yeah. I find that creepy.

The message of all those shows is a warning against passiveness. Don't be a sheep or this will happen to you.

Text Message

5

Mail-Order CRISPR Kits Allow Absolutely Anyone to Hack DNA Scientific American

CRISPR wasn't the first way to edit the genome, just like the Apple Macintosh wasn't the first computer. But there was something about CRISPR's ease of use (at least the way it was portrayed in the media) that made it seem like it would lead to this great mass democratization of gene editing–the biological equivalent of every home having a desktop computer connected to the internet. Rather quickly, there were reports of *high schoolers* doing CRISPR in biology class. Plenty of people found this inherently scary.

Pundits raced to the microphones to declare themselves the voice of temperance. *Slow down*, they declared. Get your hands off that shaker flask. Put those reagents down.

Yeah, you *high schoolers*! Go back to taking Adderall, smoking dope, and posting selfies on Instagram! Stay away from... biology class.

There was a kind of blatant obviousness about these warnings. Saying that genetics could be dangerous is like saying fire is hot. Warning that soon everyone will be able to hack the genome is like predicting, back in early caveman days, that soon everyone will have fire.

One of the things that really gets under my skin is when policy wonks go on television and say, "We need to have a *conversation* about CRISPR and where gene editing is going." Editorials are penned that urge, "It's going to have to be regulated." Threat scenarios of rogue geneticists and bioterrorists are raised. Routinely, our field is characterized as advancing with the technology *too* fast.

What irks me is any connotation that this "conversation" isn't already happening. Or that genetics is not subject to both extensive FDA regulation and Institutional Review Board oversight at *every single hospital around the world*. I've worked with the FBI's Weapons of Mass Destruction / Biological Countermeasures Unit. And I've worked with the Department of Defense's Center for Global Security Research. I was brought in specifically because our lab is a very public example of the trend that genetic engineering is no longer confined strictly to academia. Our strategy sessions focus on how to bring the safeguards that exist in academia to private labs.

So rather than waiting to the end of the book to note, "It ought to be regulated," I thought I'd address it early. Yes, it ought to–and it *is*.

I don't want to convey that the community has all the answers. But I want to assure you that the community isn't asleep at the wheel, and we aren't novices when it comes to doing the right thing when faced with tough choices.

Tough choices are common in our business. Just yesterday, at Stanford Medical Center, a patient was on his deathbed.

One of our companies is running a study of a procedure in a life-threatening illness. The patient was very ill, and our procedure hadn't saved him. But the patient knew that we have a second procedure in development; it's not in clinical trials yet, nor had we submitted the second procedure to Stanford's IRB, the university's review board. The patient and his family contacted our company's CEO, begging for the second procedure. The CEO immediately called me, because I'm one of his board directors. I said no.

It's agonizing, but it's also the law. Even though we believe that our second procedure might have helped him, the decision was clear.

This is not a rare occurrence in our field; decisions like this have to be made every day.

I guess in theory we had a chance to save the patient's life yesterday, and if we had succeeded, it would have been very newsworthy. But that didn't cross our mind.

A little history is needed.

In 1974, Stanford biochemist Paul Berg was on the verge of creating the first host cell with DNA combined from two other organisms. He was blending DNA from two viruses into a bacteria cell. His colleagues urged him to pause his work and convene a conference to discuss it. Berg listened to his peers. The now-legendary Asilomar Conference was held in Monterey, California, in February 1975. One hundred forty people showed up, most of them scientists, but also lawyers and ethicists. Biohazard principles were established. Risk-assessment protocols were implemented, and risk-containment strategies were endorsed. The following year, Genentech was founded to usher in the era of recombinant genetic engineering. Berg would go on to share the Nobel Prize in 1980.

The Asilomar Conference happened shortly after President Richard Nixon resigned because of the Watergate scandal. The scientists understood that secret, government-funded activity inherently bred distrust. Going to conferences and openly talking about their work has been a fundamental principle of science before and after.

Yes, high school science classes use gene-editing CRISPR kits today. They also use chemical agents. They also program computer code. They are about as dangerous as Mentos dropped into a Coke bottle.

A few years ago, scientists in Canada showed they could mail-order the DNA of the horsepox virus. They couldn't order the entire DNA–that would have been caught by security databases–but they ordered it in small chunks, and then recombined it in their lab. The horsepox is not dangerous, but in theory, they could continue editing it into the smallpox virus. (It wouldn't have been easy; the two poxes are 34,000 base pairs different in length.) But they wanted to reveal the vulnerability, and they did. Though the industry rapidly caught up.

It's important to understand that when scientists want to insert some genetic code, they almost never create that new code string from scratch. They order these chunks of code online from a DNA synthesis company. It's incredibly easy–you just type in the letter sequence you want, then hit "Add to Cart."

But every licensed DNA synthesis company checks an order against the U.S. Federal Select Agent Program, the Functional Genomic and Computational Assessment of Threat (Fun GCAT) program, and the DARPA Pandemic Prevention Platform. They compare the requested DNA strands–even small chunks as few as two hundred nucleotides–against databases of all known pathogens. The biggest companies also run a computation on what the chunks could be recombined into. They also share order requests between companies so that someone can't get around the screening by ordering from multiple companies.

In a similar way, we were all disturbed when the rogue CRISPR-Baby scientist, He Jiankui, presented his work on the two girls he tried to genetically immunize from their father's HIV infection. To pull it off, He had lied to his hospital, and may or may not have lied to the parents. The Chinese government swiftly incarcerated the scientist for three years, and dramatically raised the financial penalty (to about $800,000) for doing work without proper permission. As well, an international whistleblower system was put in place. He Jiankui had spoken to a half-dozen American scientists about his work over the years; every one of them had tried to dissuade him. But because Jiankui was in China, they didn't have a way to report him. Now they do.

Over the course of this book, we'll get into a better understanding of CRISPR, and why it's a lot harder to use on a multicellular organism than commonly portrayed.

The most common nightmare scenario offered is not a rogue academic. It's that some high schooler will accidentally create a mutant bacteria and release it into the world. (High school CRISPR kits edit an *E. coli* bacteria genome.) What people don't realize is that all bacteria are constantly mutating already. Bacteria have a way to swap DNA on their own–and they don't need CRISPR to do it. Every time we use antibiotics, we fundamentally force bacteria to mutate to try to escape. Of the trillion species of bacteria, fewer than a hundred are pathogenic to humans. Introducing CRISPR into this mad scramble is like bringing a Philadelphia Eagles jersey to a football game at Lincoln Financial Field. It does absolutely nothing, because every fan *already* has an Eagles jersey–just like every bacteria *already* has CRISPR in it. (CRISPR comes from bacteria.)

I think it's time to bring into this discussion the CEO of Twist Bioscience, Emily Leproust. Twist Bioscience is a leading DNA synthesis company. I called Emily about the biosecurity risks.

"Around 1 percent of our orders are flagged," Emily explained. "Most of the time, they are science labs who know what they're doing and have all the paperwork. Some of the time, they had no idea that what they ordered could be combined in a dangerous way, and they're really grateful. And some of the time, they knew it was in the gray area, but they didn't know all the paperwork steps and appreciate the lesson."

She said, "If someone ordered SARS from us, even a part of it, we'd catch it no problem."

"I *am* concerned with the risks," Emily added. "There is potential for massive loss of life. But this risk isn't from some postdoc or high schooler. The largest risk is from nature."

And this is the real point. Nature is always evolving. Especially at the viral level. Nature slows down for no one.

Emily put a button on this. "Nature is the greatest bioterrorist," she said. "The biggest losses of life have been from nature."

Mankind's technological achievements over nature, throughout history, instilled a deeply fundamental bias–that the eon of being at the mercy of nature was behind us. That if we could span bridges over rivers, and land on the moon, and take a photo of anywhere on Earth from satellites…*no way* could mere nature *actually* threaten us. For most of this great technological age, nature retreated and mankind advanced. It was only our fellow man we had to fear.

And so when it came to genetic advances, we fretted over what our fellow man might do with this new wizardry. And we took our eye off the humble fact that nature is always mutating and selecting…until one day, mankind was no longer advancing. Mankind was hiding in its homes, trying to avoid the invisible menace.

All along, we humans had been inadvertently homogenizing the planet. Most pathogens, in the past, stayed in their ecosystem; they either couldn't survive outside it, or nothing carried it afar. But as we remade the world into one megasystem, we created a bridge for far more pathogens to spread.

In the early years of IndieBio, the Zika outbreak was emerging in South America. We funded several companies to fight Zika and whatever new infectious disease might emerge. One founder went to Brazil; in one lab, he found thousands of untested samples, and in the hospital, babies with microcephaly. Working with DARPA and the Gates Foundation, they invented a small clear chip, just the size of a thumb drive, which worked with a smartphone camera to get a diagnosis–just like a $20,000 lab device.

Another amazing company became the world leader in bioprinting vascularized human tissue. While at IndieBio, they printed and grew panels of miniature human lymph nodes, little immune systems. Then they inoculated these lymph nodes with the Zika virus, and the lymph nodes rapidly–naturally–developed antibodies. By extracting and sequencing those antibodies, they had an extremely fast way to deploy antibodies against Zika, and whatever came next.

Both were astonishing technologies. But even during the Zika scare, they couldn't get more funding from VCs to pursue infectious diseases. Fighting pandemics was not considered an investable market, simply because nobody knew when or where the next epidemic would happen. It could be years or decades. Both companies survived, and they did get funded–but to use their tech a different way. (And both went back to their original work when the pandemic surged.)

The story of "We Weren't Ready" didn't begin with a shortage of N95 masks and ventilators. It began decades prior, when we tuned out SARS, H1N1, MERS, Ebola, yellow fever, dengue, CHIKV, cholera, and meningitis, and we grew false confidence that maladies like that didn't happen to good people like us. So the funding wasn't there, from private or public sources, to support technology that could have been ready to go, well in advance. We believed in mankind's invulnerability, mankind's dominion over the wild.

6

The Mystery of Vanishing Honeybees Is Still Not Definitively Solved Science News

This is a chapter about saving the bees, and my first thought was to have an AI write it.

It took about a minute to plug in keywords and about three minutes for the AI to write the story on the phenomenon of bee colony collapse.

My expectations weren't high. I figured using an AI might simulate what someone might learn through reading about bees on the internet for an hour. The AI located and sucked in thirty-six different sources, from Environmental Protection Agency briefings to the *Los Angeles Times* to legal blogs and university publications.

The AI also allowed me to use a slider bar to increase or decrease the amount of text it rewrote from the original sources. It defaulted to 59 percent, meaning 41 percent was still fully cribbed from the sources, and then, as I ramped up its rewriting to the maximum 75 percent, it operated like a high schooler, trying to escape copyright violations while still making sense. It warned me, "Very high rewrite-levels are usually neither needed nor advised, due to degrading text coherence."

I'm not trying to make fun of AI prose. That would be too easy a joke. In fact, to honor this plebeian machine, I'll drop some of its sentences in here as I go. I was actually much more interested in the core content it extracted. And how that "Book Knowledge" compared to the wisdom and understanding we got from running a $250,000 experiment in *solving* colony collapse.

But it really wasn't a fair comparison–the AI created a mere 825-word summary. Our experiment has been running for a year.

So, I paid a high school sophomore (my daughter) $15 an hour to read all thirty-six original sources. Then I interviewed her and compared our wisdom to her (probably temporary) knowledge.

There is really nothing boring about bees. From their furry-robot look, to the radical way they fly, to how they communicate the location of the best flowers with their wiggle dance. Bees can even do a little math. But in 2006, all of a sudden bee colonies started collapsing in huge numbers.

> "Bret Adee, co-owner of Adee Honey Farms, said that when it came time to pollinate his almonds, nearly 55 percent of his 80,000 Honey bee colonies (two billion bees) disappeared."
>
> –AI-Writer

For over a decade, "Save the Bees!" stories ran. Beekeepers gave speeches and went on television. We were all asked to imagine a future without bees. In every story was a clear message that humans had ruined the ecology for bees, harmed their ecosystem any number of ways. We needed to save bees *for our own sake.* Half of agriculture relies on them.

All of that is accurate, but book knowledge also engenders this kind of intellectual space-time distortion, where two perceptual phenomena tend to happen. These plague futurism generally, especially futurism uninformed by direct experimentation. The first is what Arvind and I nicknamed the "Warp-Speed Impulse." The second is a "Black Hole Effect."

Warp-Speed Impulse entails this tendency to immediately extrapolate into the far-off dystopian future. In the case of bees, dire predictions about a world without bees. We see this warp-speed impulse all the time. When CRISPR was invented, almost immediately people were talking about the future horror of designer babies. The driverless car led people to impulsively jump ahead to a future without jobs. A single blood-sharing study in mice, in which the older mouse grew younger, led everyone to race ahead to contemplating a society where everyone lives to 150. Not all of it is dystopian; when checkpoint inhibitor immunotherapy was invented, people started predicting a future without cancer.

It's not that these thought exercises are necessarily wrong. It's just that the future doesn't tend to obey these statistical extrapolations. The future is not a linear extrapolation, nor is it an exponential extrapolation. There's a lot more to it (which will be a recurring topic in this book).

The Black Hole Effect is often a subtle or an implicit intellectual phenomenon. Real black holes tend to suck all surrounding matter into their abyss, whether it's nebulas or planets or stars. In the metaphorical version, it's the tendency for a world problem (such as bee colony collapse) to suck into its abyss all the systems it touches. Laws, economic systems, entire global industries, ideologies, even the fundamental nature of human behavior. These are often encased in the phrase "the *real* problem."

ME: After reading all these articles, what do you think is the issue in bee colony collapse?

FIFTEEN-YEAR-OLD: There are a lot of contributors. But the real problem is our system of industrial agriculture. I read how our crops and our bees have no genetic diversity. And the financial system. Which forces agriculture into growing only one crop and abusing pesticides.

When the Black Hole Effect gets going, capitalism itself is often to blame. Or bad government policies that fail to properly incentivize capitalism to do the right thing. Or bad campaign finance laws to prevent greenbacks from influencing wise regulations. Which may all be partly true, but when you're trying to solve the problem on the ground, with bees, you don't get to think that way.

At IndieBio, we take a more direct route. We fight capitalism with better capitalism.

Where the Black Hole Effect tends to look outward for problems (and solutions) in ever-widening scope, we look for problems (and solutions) in an ever-narrowing scope.

We found a team of scientists in Argentina who had written an unusual science paper on the complex interactions between plants and bees right on the flower. They postulated that the solution to bee colony collapse would be *naturally already present*, just in too low a quantity. The most common pesticide–the one that was hurting bees–basically turns every sprayed crop into a tobacco plant. The nicotine in a tobacco plant is a neurotoxin; it doesn't kill a bee so much as it compromises their immune system. If you think about it, pesticides are designed to kill insects. Bees are insects.

The scientists were confident that in the delicate petals of a flowering plant, there would be *natural* elements that would benefit a bee's immune system. Fundamentally, bees and plants would *share* beneficial molecules. They found several of these. One was a plant hormone that helps protect plants during winter months; passed on to bees, it also helps bees through the winter. Another was a key plant amino acid that helps come to the rescue when a plant is scratched or torn or broken. Passed on to bees, it helps protect bees from parasites–such as the mites that carry the deformed-wing virus that was wiping out bee colonies.

> "Bee colonies are often found to be infested with pathogens and parasites, and research into the interactions of all possible causative substances has proven to be a challenge for bee scientists."
>
> –AI-Writer

We truly had no real idea if there was a viable business there. But we loved the science. And the scientists, Pedro and Agustín, had teamed up with a young entrepreneur, Matías. We gave them $250,000 to get started in California. Their natural additive to hive honey worked wonderfully on the bees, boosting their immune system and preventing colony collapse. And their plant medicine (they actually called it "chimichurri") was affordable for beekeepers.

Here's where the Warp-Speed Impulse comes back into play. You click on news stories like this every day: "Scientists Save the Bees–with Plants!" And you read about their important scientific breakthrough. You warp-speed ahead, imagining venture capital pouring in money, imagining every beekeeper using it, imagining the entrepreneurs making a tidy sum. Their solution takes over the world as fast as Facebook did, and it's a world *with* bees.

But sadly–that is not what you learn when you actually run the experiment. Instead, you learn that most beekeepers are far from scientists, and not really experts in bees at all. Beekeeping hadn't changed much in a hundred years. They didn't want to risk their precious bees with exposure to anything new. Certainly not something from a startup.

The beekeepers didn't want the chimichurri.

> "Beekeepers need to set up their colonies are [*sic*] suitable places that allow access to suitable food sites and are far from areas where pesticides are used (honeybees can feed up to 5 km from their colonies)."
>
> –AI-Writer

So our little three-person company had to *become* beekeepers and compete. And as long as they were going into the beekeeping business, they might as well be scientific about it. They put tiny cameras on every flowering plant in their test garden, and they put tiny colored tags on the bees' legs. They tracked every flight path, cross-referencing honey production and raspberry fruit size. They even tracked how much nectar each berry secreted; it's this nectar that attracts a fungus and causes berries to mold. They discovered that beekeepers, for a hundred years, had been placing their hives in the wrong locations. The more times a bee visits a flower, the bigger and healthier the fruit or nut will be. By shortening the routes that bees need to fly, they can hit each flower more; you get healthier crops *and* healthier bees, because they eat better.

They also found that when bees come out of a hive in a new orchard for the first time, they don't know what to do. They don't just automatically go pollinate what the farmer has paid the beekeeper to pollinate. A hive of fifty thousand bees spends a lot of time experimenting to figure out where the protein-rich pollen they need for their diet is. Our scientists figured out a way to make bees go straight for their intended targets, by putting yet more plant hormones in their chimichurri. One mix primed them to go to almond flowers; another to blueberries, et cetera.

Our company, Beeflow, just finished their first full spring of testing. They didn't have to get rid of capitalism after all; in fact, their big investor is a Wall Street hedge fund. They also didn't have to get rid of industrial agriculture. Their testing was with the world's largest berry farmer and the third-largest almond grower in the United States.

We just had to get rid of bad science and replace it with better science. Our investment thesis is simple: Whoever has the best biology wins.

In every field tested, the bees suddenly thrived, in ways the growers hadn't seen in decades. The bees visited each flower 2.6 times more than before. A couple months later, the berries exploded on the branches. Normally on a berry farm, only half the berries are sellable in grocery stores. With Beeflow on the job, that leapt to 80 percent.

And colony collapse can be a thing of the past.

Last week Matías, the young entrepreneur who led this company, had a long meeting with the president of Argentina, Mauricio Macri. Their photo was in the newspapers in Latin America. But artificial intelligence apparently hasn't discovered Matías yet.

> "The problem of vanishing honeybees is still not solved. A survey of 4,794 beekeepers from all U.S. states and territories said 40% of their hives, also called colonies, died unexpectedly during 2018. That's up from 33% in 2017."
>
> –AI-Writer

I don't know why the AI-Writer doesn't know about Matías. I also don't know why the AI that Beeflow created to track bee routes doesn't hack the Wi-Fi and touch base with the AI-Writer now and then. *Hey, dude, give me a little credit here.* Maybe it's not "the future" yet.

By running the experiment, we learned a lot about bees. But we also learned how the world *thinks* about global problems, and we identified some of the common pitfalls. One of the most important lessons of all was this: Pedro never learned exactly what the "real" cause of sudden colony collapse was. That in fact is still a mystery.

ME: But it turned out Pedro didn't need to know what precisely caused it. He just needed a better way to make bees healthier.

FIFTEEN-YEAR-OLD: So when I eat raspberries now, I'm doing good for the bees?

ME: Next spring we hope Beeflow is at every field in California. We'll see.

FIFTEEN-YEAR-OLD: "We'll see"? What do you mean? And why just California? Why wouldn't farmers do it in every field in the world?

ME: Yeah. I guess I'm learning that it's harder for the world to change its habits than I hoped.

7

What Colors Are This Dress? BuzzFeed

On an island in the Inner Hebrides of Scotland, a woman named Cecilia Bleasdale sent an overexposed picture to her daughter of the dress she was going to wear at her daughter's wedding. It was a royal blue "Lace Bodycon Dress" from the retailer Roman Originals. The bride-to-be posted the image online, and everyone began arguing over the colors they were seeing. BuzzFeed picked it up and ran a poll. Within a few weeks, hundreds of millions of people had seen it online. Two-thirds of them said the dress was white and gold, not blue and black. It was the dress that broke the internet.

And it was quite a parable for our divided times. If we couldn't even agree on what colors we were seeing, then how could we ever agree on taxes, or climate change, or immigration?

It was shocking that something so simple as color perception could contain such dramatic observation bias.

It wasn't that we saw different colors, per se. Seeing blue is not the problem. Take any cell of your body, peer into the DNA, and go to chromosomal location 7q32.1. That's where the blue opsin gene is located; it encodes a light-sensitive protein, cyanolabe, that changes shape when struck with blue-spectrum light waves.

But seeing color is a small part of observing. Observing happens in the brain. And our brains have to make sense of the color signal. The most satisfying explanation is that our brain has gotten very good at smoothing out colors, for continuity's sake, amidst constant lighting differences as objects pass in and out of shadows and against contrasting background colors. When we look at "the Dress," we're compensating for what we think the lighting is doing. Like those brightness and contrast slidebars on our photo apps, we're instantly converting the image to what we *think* it's supposed to look like.

Our brains trained themselves to do this, and apparently what the sky looked like, when we were children, influenced our compensation for shadows on blue.

Blue is soaked through my memories, and every shade of blue is a different primary remembrance. Periwinkle is the tint of my daughter's eyes. The ultramarine of sunset through our living room window always makes my wife recall Maxfield Parrish's paintings, heavy on the varnish. The most beautiful turquoise I ever saw was the baywater inside Shackleford Banks, in North Carolina, where the wild horses stood belly deep in the shoals a half mile from shore to cool off. My brother's coastal home is there.

So I want to ask this question. It sounds innocuous, almost childlike. What would be the Perfect Blue? If you could make *any* blue, what would it be?

Would it be royal blue? Misty blue? The blue of your childhood sky?

Maybe it would be the blue of Earth as seen from the moon.

Plato said that nobody had ever seen a perfect circle, or a perfectly straight line—but everyone has the idea of a perfect circle and straight line in their mind. And by his logic the Perfect Blue can be only in our minds, never matched by a color in the real world.

I'm going to take a guess that no other venture capital firm in the world sits around debating the question of what would be a Perfect Blue. They probably don't think of blue as a technology at all. A Perfect Blue is not an investment thesis on Sand Hill Road, where the mandarins of venture capital ply their trade. Federated artificial intelligence systems might be more like it (though federated learning is a thesis of ours, too). If they had to answer the question of what makes a blue "perfect," they'd probably ask their kid.

But talking about a Perfect Blue is exactly the sort of thing that happens at our firm. We don't plan to do it. It just happens in the course of conversation. And it comes up for us because we study history a lot. If you read your history, you realize: Humankind has been on a quest for the Perfect Blue for thousands of years. And they will continue to search for it in the coming thousands of years.

For most of history, blue wasn't called blue–linguistically, it was just lumped in with grays. People didn't give it a name. In the *Odyssey*, Homer described the Aegean seas as "wine dark." He never once used any word such as "blue." In the Bible, there is no mention of blue. The Mahabharata describes the oceans endlessly but never describes their color. Red, green, and yellow all get a lot of usage in the ancient texts, as do black and white, and shades of darkness.

The reason the word "blue" was rarely used was that mankind wasn't turning many things blue.

The technology of making things blue was centuries and millennia behind other colors. To get blue, you had to grind stones into a powder and isolate the lazurite. And you had to get the rocks all the way from Afghanistan, which was a long way from everywhere. Those rocks were heavy. The Egyptians used it for royal tombs. Blue jewelry was sold, but it wasn't that precious. Blue minerals could be ground to powder. Fabric was dyed primarily with woad, a plant that cast a light gray-blue. In western culture, blue became treasured only in the 1100s, when cobalt was mixed into the stained glass windows of the Basilica of Saint-Denis in France. The light it cast in the church attracted people from everywhere. When ultramarine paint was created (more grinding of the rocks from Afghanistan), the Roman Catholic Church declared anyone painting the Virgin Mary needed to drape her in blue robes. Good ultramarine was more expensive than gold. During the Renaissance, using lots of ultramarine was mandated in artists' contracts. Blue became a highly desired color. To satisfy the desire for blue apparel, trade routes with India were established to import plant indigo. Both Germany and England banned indigo, to protect their inferior woad industries until they could set up their own trade routes.

The history of indigo is a history of exploitation of the global south by wealthier countries. In the 1800s, the British convinced Bengali farmers to grow indigo instead of food; they loaned the farmers money, but paid the farmers only 2.5 percent of the market value of the indigo, so these farmers lived in permanent debt, which was inherited by their children. Revolts were suppressed ruthlessly. By the end of the century, chemical indigo had been invented, and it was heralded as an incredible achievement. From then on, all the world's blue jeans were treated with chemical indigo.

So with this in mind, knowing all the powers that humankind has over science today, what would be the Perfect Blue? Take a moment to consider it.

What you soon realize is, it's *not about the color.*

Oh, the color matters–you need some choice there still–but if you're looking for the Perfect Blue, it's not the hue. Don't judge blue by its color.

Here's some rules for the Perfect Blue. Some criteria. See if you disagree with any of these:

1. It's got to have all the color characteristics of the existing blues we love.
2. It should be affordable, because it's important that this Perfect Blue be available to everyone, not just the rich.
3. Whoever works in these blue factories, or blue fields, should be paid a livable wage. And have health care.
4. Their children shouldn't have to work in those factories or fields.
5. If it's going into our food, this blue shouldn't be a synthesized *chemical*, made from petroleum–it should be *food.* Like good old protein, carbs, or fats.

6. It certainly shouldn't be a chemical that is connected to health problems, and most of all not brain problems. We should be able to eat a whole bowl of it, like a bowl of blueberries, and be *healthier* for eating it.
7. Whoever owns these factories of blue shouldn't be using other hazardous chemicals in the process. If it's coloring our clothes, then our clothes shouldn't have any chemicals in them.
8. It shouldn't be contributing to climate change.
9. No country should be marginalized or exploited in the process, forced into providing cheap labor or precious resources at bargain prices.
10. It should still remind us of the skies of our childhood.

Okay, there's the 10 Rules for Blue we invented over coffee. I mean, it's just a color...so this is sort of abstract at this point...but I think it's fair to say, this is how we'd prefer to make things. If we had the chance to invent the means of production, then we'd all agree, yeah, this is better.

So if you're on humankind's quest for a Perfect Blue, now you're thinking harder about how things are made, and the impact on people and their governments in the process. Throughout history, blues were either insanely expensive (you had to be King Tut to afford them), or someone in the supply chain was getting royally screwed.

And you can't go around saying, "We should just stop liking blue. If blue is so bad for _______ (the environment, worker health, inequality), then give up blue." Humankind's desire for blue is deeply seated in our culture. It's also deep in our brains, quite literally. Dozens of brain regions that never see sunlight nonetheless have blue-light-sensitive opsin genes in them. They control our circadian rhythms and, even more important, up-regulate astrocytes that protect our brains from damage. One of the coolest new fields of medicine uses blue light as a drug to act on the brain, triggering these blue genes and the cascade of genes they up-regulate. We evolved to have these, because blue was the color of our skies and oceans.

So, no, we're not giving up blue. Finding Perfect Blues is the way ahead.

Let's look at the Rules for Blue and apply them to denim blue jeans.

To make chemical indigo, so much cyanide and formaldehyde is used that its production is banned in the United States and Europe. Then, to make the indigo soluble in water for dyeing, more nasty chemicals are used. Wastewater-treatment facilities can't filter some of the chemicals out. Workers have significant health problems. Impending regulation in China is pushing the manufacturing into Mongolia. Every time we wash our jeans, trace amounts of cancerous agents bleed into our lakes and oceans. Every denim company knows this. They all have official sustainability reports, but none ever mentions indigo. It's the denim industry's dirty secret. Even if you use plant-based indigo, the pigment is not water soluble, so tons of chemicals are used. (It was always hazardous, even before chemical indigo was created.)

The denim industry is booming nonetheless. It's a $100 billion global industry. Denim is popular all around the world. Indigo has something that regular blue dyes don't have; it binds to white yarn. Inside the yarn of your denim jeans, the cotton is white; only on the outside is it blue. As you wear your jeans, the indigo slowly breaks free and white starts to show through. This signature quality allows something rare in this world: for everybody's pair of jeans to be unique. People love their blue jeans.

Wouldn't it be great if we could make a Perfect Blue for blue jeans?

Let's also apply the Rules of Blue to food dyes.

You may not think we color a lot of food blue. Except for popsicles, candy, and kids' yogurt. Oh, and ice cream. And cereals with blue things in them. But we actually use blue–mixed with yellow and red–to make foods chocolaty brown. So a whole lot of things that are brown in the grocery store actually use blue food dyes. The FDA approved our two blue food dyes in 1931. I don't know if they cause ADHD–but the American Psychiatric Association didn't recognize ADHD as a condition until 1962. What I do know is that our blue food dyes *aren't* food. One is a chemical made from petroleum, the other is a synthetic indigo (discussed previously). Maybe I shouldn't freak out–I mean, 99 percent of beers have pesticides in them, and we're all okay. But I think it would be kinda cool if blue food dye (and chocolate-brown foods) were made with actual *food*.

As long as we're at it—as long as we're imagining a Perfect Blue—give me a blue with antioxidants and supernutrient properties.

You may be thinking, about now, "Nobody but those weirdos at IndieBio are looking for a Perfect Blue." And sometimes we worry about that, too. But luckily for us, we are not alone. There are scientists who are just as weird as us. Scientists who are sorta like Maxfield Parrish, sitting in their labs thinking, "How can I make a better blue?" And rather than wanting to paint a landscape with their blue, they had the creativity to go in another direction—to imagine a world with blue jeans we could be proud of. To imagine blue and brown food that was actually food. Chemical-free blues. And when we find these scientists, we may be the only venture capital firm that really understands the cultural significance of their inventions.

So, yes, blue is the color of our memories. But also, *blue is a technology*.

Everyone has this implicit, unspoken, but pervasive notion that technology will bring us *new* things, wonders we've never seen before. We'll have clothes dressers that do the laundry for us, right in the drawer—or maybe we'll be able to text each other a design for an electric submarine with a thought.

But, you know, before we make cows fly, technology can do a lot better just reinventing the things we already love. The things we always treasured. The desires that always drove our quests.

We will get to the wild and weird side of the future over the course of the book. But here, as we start out on our journey, we want to stick with ordinary stuff that we tend to take for granted. Like the blues, sweet blues.

---

* Everything in this chapter was written while listening to binaural music at 40 Hz. It's kind of a white noise, though it also sounds like alien spaceships in old television shows. But it's been shown to operate on the deep brain similarly to blue light (also at 40Hz), increasing gamma wave activity and up-regulating anti-inflammatory, neuroprotective mechanisms in our neurons. I feel great.

8:48 PM
92%
Messages
Arvind
Details
something weird happened when I wrote The Rules of Blue chapter...
what happened
I was listening to binaural beats at 40Hz to increase my brain's Gamma waves
then of course something weird happened
I was all peaceful and meditative...but then I got really angry...
interesting
thinking about how Tinctorium was having such a hard time raising money when it's such an incredibly good company

VCs should be throwing money at them...while all these other dumb companies in the valley rake in tens of millions

I just got really angry. and then I noticed -- I had turned off the music without realizing it

hahaha

Do you get mad?

no

really?

I think about being a Seed/A investor, where I can help these startups get to the next level

leave IndieBio?

I don't know. So torn.

Text Message

8

10 James Bond Villains, Ranked by How Evil Their Plans Were Screen Rant

I'm standing here on the side of the road near Bogatynia, Poland, looking through a black wire fence at the biggest hole in the ground I've ever seen. I'm watching the open-pit Turow coal mine in full operation. The trench is about three miles across and wide; in the distance, I can see the boiler stacks and cooling towers of the power plant that burns the coal for electricity. Massive graders and dozers are shaping the gray-and-ocher stepped benches that descend into the darker earth. Conveyor belts and mining trucks are ferrying dirt and coal this way and that. The air above the mine seems to shimmer. I don't see blasting, but orange-painted excavators bigger than houses are scraping out craters, and crushing machines are breaking up the bounty into chunks. From my vista point, it's impossible to see how deep it goes.

I wanted to look right into the bottomless pit of the enemy, down into its dark hole. And then I am shooed away by a security guard. I stop taking pictures and climb back into our car.

The Turow power plant is ranked as the number one worst greenhouse gas polluter in Poland; it's not quite the biggest mine, but at Turow they mine and burn brown coal, lignite, which is the dirtiest of coals, rich in volatile compounds and ash. Even China, which burns half the world's coal, has severe restrictions on brown coal. In the winters here, the snow covering the town is black.

Despite the availability of cheaper renewable sources, coal is still Poland's go-to power source; they get 80 percent of their energy from it. Elsewhere in Europe, they've at least stopped building new coal boilers. But at the Turow mine, the power station is in the midst of constructing their eleventh boiler, to come online next year. I can see it going up in the distance. The country is increasing coal output by 10 percent.

History will show the first half of the twenty-first century was the end of coal's long tenure. Poland has pledged that they will get rid of half their coal plants by the year 2040, and the European Union has declared no more coal at all after 2050. Yet old habits die hard. This coal seam at Bogatynia has been mined for 279 years. And despite people complaining that the world is changing too fast, there are way too many places, like the black pit of Turow, that are not changing at all.

Bond films are always stories of technologies pitted against each other. Early in the films, 007 picks up a few cool gadgets. He starts following the trail, from one heist or assassination to another, eventually uncovering an evil villain who has gained control of another fantastical technology, usually much larger in scale. Nuclear beams, decoding devices, bioweapons, satellite laser weapons, extraterrestrial nerve agents, electromagnetic pulse weapons, and mass surveillance systems. The villain targets our happy way of life and figures he's going to disrupt it with a calamity. Magically, in the middle of the dustup, 007's small handheld technology proves essential in defeating the plot to remake the world order with megascale technology. The films aren't antitechnology; rather, they're populist. In the Bond bible, big tech is synonymous with evil; small tech is cool.

But I'd like to see a Bond villain who wasn't out to change the world. I'd like to see a Bond villain who realized the greatest power and damage is caused by *preventing the world from changing.* Now, that might give the Bond series a tinge of Rod Serling–era *Twilight Zone*, but that wouldn't be a bad thing. It starts with Bond at home doing his laundry, when his washing machine breaks down. He goes to the local Euronics shop to buy a new one, and discovers there *aren't* any new ones, just old ones. He stops by the Aston Martin showroom, and there's no new model this year. Grumpy, 007 heads to MI5 to get some gear, but Q hands him the same set of gear he had in the last film. At first, all of London is kinda excited by this giant pause in technological progress. An assassination and a heist puts Bond on the trail of a secret international organization called *Inertia*. A double agent confesses on her deathbed that *Inertia* is out to prevent the spread of new technologies. 007 infiltrates the Bank of England and learns that *Inertia* has taken control of the financial system and its operatives are already planted in Westminster. The conspiracy is deep.

Inertia is the resistance to change. And while maybe *Inertia* wouldn't make the best Bond film after all, it is the villain we fight day after day.

The Turow coal mine is just one of many heists that illustrate the role that *Inertia* plays in our real world. The only explanation for *increasing* coal production in the year 2019, with the climate crisis in full swing, is Isaac Newton's First Law of Motion. The bigger the mess, the *easier* it is to just keep going the same way we've always done it. And the Turow mine is a *big* mess.

I follow the road several miles around the coal pit to visit the town of Bogatynia, which is nestled right up against the mine's fifteen-foot concrete wall. The power plant welcomes me to town. Bogatynia is a paradox. Its architecture is all Soviet-style unadorned buildings, many of them unpainted concrete. The sight of them is a shock compared to the gorgeous Italian palazzo and Gothic Revival architecture in the German towns just fifteen minutes across the border. There are a few fading villas and a few new condo complexes. The thing is, Bogatynia is not poor–it's actually one of the richest cities in Poland. The coal mining union is very powerful here politically, and the wages in Bogatynia are high; miners make about 20 percent more than most workers. The Polish word *bogaty* even describes someone wealthy. I learn that Poland is sinking about a billion dollars into the new boiler for the power plant, and that when the wind doesn't blow, the skies get dark gray.

I'd like to think that Poland is an exception in our world. But in the same two decades that affordable clean power has been brought online, global coal capacity has nearly *doubled.* And at no time in the last two decades did coal violate Newton's Law of Inertia. It didn't stop even for one year. Every year since 2000, coal capacity has grown. According to the World Bank, the world gets 40 percent of its electricity today from coal. Back in the 1970s, it was only 30 percent.

Most people don't know this; most people would say the dependence on coal *must* have gone down a lot in the last two decades, and even more over the last half century. But inertia is kind of invisible–unless you decide to look it in the face. When you do, it's too vast to even get into the photo frame.

Fifty years of an environmental movement has been no match for the insidious villain of *Inertia.* Even the United Arab Emirates, which has both all the oil it could ever need (if it wants to stick to fossil fuels) *and* so much sunshine and open land for solar power (if it doesn't), is building coal power plants now.

At the local discount store along Pocztowa Street, I go in hoping to buy a bag of coal, on a whim, but they don't sell it. I try to ask where to buy it, but my question is confusing, even for the clerk who speaks English.

The ultimate irony of coal is lost here in Bogatynia. All the coal on Earth, including the coal being mined just a quarter mile from here, owes itself to a brief imbalance in Earth's ecosystem that ended 300 million years ago, the Carboniferous Period. During that time, plant life was evolving. For the first time, plants had the genetic ability to create the fiber lignin to strengthen their cell walls. This allowed them to grow magically tall, and the more they grew, the more the plants next to them had to reach even higher to find sunlight. But all these plants didn't have root structures adequate to their height, so they fell over in the swamps, piling on top of each other.

During the Carboniferous Period, trees had lignin, but bacteria and fungi had not yet evolved to feed on lignin. Today, when a tree falls, bacteria and fungi break it down slowly, releasing all the carbon back to the planet's ecosystem. But during the Carboniferous, fallen trees didn't break down—they held on to their carbon and, with time, became coal. Coal can be up to 95 percent carbon, but brown coal—lignite—is about half carbon. Lignite gets its name from the fiber that protected its carbon stores. The Carboniferous Period ended with climate change.

So you could argue that coal really had two active periods on Earth. It took 60 million years of trees falling over to make our coal reserves, and then 300 million years later, we burned it up—in just 150 years. Climate crises will bookend both chapters.

Coal is a big business, but nobody loves coal itself; amazingly, inertia persists even though coal is largely unprofitable. Four out of ten coal power plants around the world are running at a loss, according to the London-based think tank Carbon Tracker. Half the coal industry can't even say they are in it for the profits. They keep operating only to pay down the bonds they sold years ago, when they borrowed the money to build the new billion-dollar boiler at the power plant. This is *Inertia*'s ray gun. Its nerve agent. They make the money cheap to borrow, and then that debt turns the borrower into a weapon *against* change. As long as the world doesn't change, the bonds are paid off.

Today, lignin is under our control; we can engineer both plants' ability to make it, or not, and fungi's production of enzymes that break it down (or not). We can grow gargantuan fast-growing fern forests that sequester carbon and don't release it, even when the tree is felled—or we could deacidify the oceans with carbon-sucking marine plants. Or maybe we don't do that; maybe we hold off on that for a while. Maybe we just *stop burning the coal.* But having a solution is not the same as implementing the solution. Right now, *Inertia* is still winning.

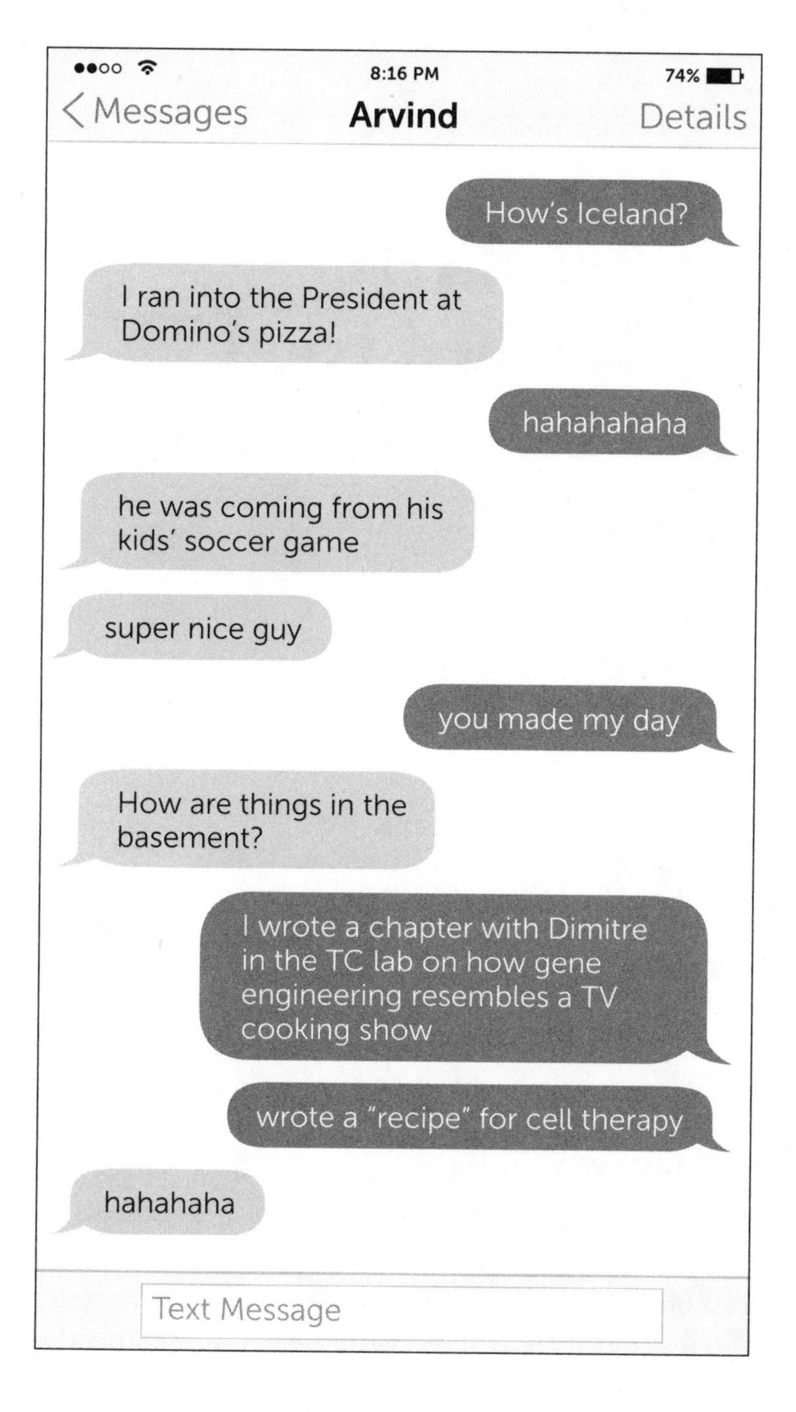
●●○○
8:16 PM
74%
Messages
Arvind
Details
How's Iceland?
I ran into the President at Domino's pizza!
hahahahaha
he was coming from his kids' soccer game
super nice guy
you made my day
How are things in the basement?
I wrote a chapter with Dimitre in the TC lab on how gene engineering resembles a TV cooking show
wrote a "recipe" for cell therapy
hahahaha
Text Message

9

What Are the 10 Most-Watched Food TV Shows? You Might Be Surprised PopSugar

Genetic engineering is a lot like a television cooking show. But it's also not remotely like a television cooking show. To explore the similarities and dissimilarities, I decided to grab the nearest geneticist, go into our lab, and perform the exact steps of editing someone's genome.

The nearest geneticist is usually the guy who sits right next to me at work, Cody. But he was out today, so I grabbed the nearest geneticist across the aisle of desks. This was Dimitre Simeonov. He's thirty-two, and had come to IndieBio from UC San Francisco about eight months before. He had been at UCSF for six years, where he became a leading expert in several aspects of genetic engineering, which we'll get into. Born in Bulgaria, he came to the United States at the age of eight. Game for the shenanigan, Dimitre takes me into the tissue engineering lab.

The dish we're going to prepare today is called "A Better Way to Do

Kidney Transplants." Dimitre would prefer to call it *Immune Regulation via CRISPR knock-ins of IL10 to T-Regulatory Cells.* Which just isn't as catchy.

The challenge of kidney transplants is that the patient's immune system can reject the donor kidney. The drugs used (there are two available) suppress the immune system, but do so all over the body, and so the patient is now at very high risk of infection *and* cancer. One of the two drugs is designated a Group 1 carcinogen by the World Health Organization. These risks mean tons of patients aren't eligible for a kidney transplant.

Dimitre's got a better way to do it–without using drugs at all. Or maybe a better way to say it is that he's going to put the drug factory inside the patient's body, so the patient's cells make their own natural drugs. The drug he wants them to make is a natural drug, IL10; the body already makes it but Dimitre wants it to make *more,* to prevent kidney rejection.

It's amazing–but Dimitre says a lot of scientists are creating recipes of this sort. He also makes it clear that there are many other ways to do it; this recipe is just *his* way.

Primary Ingredients

Standard blood draw from patient (450 mL)

Double-stranded DNA repair template (ordered online, about $500)

Use the code for the IL10 cytokine. Cut and paste the code from the UC Santa Cruz gene browser. It's 1,629 letters–As, Cs, Ts, and Gs–that begins with the sequence ACACATCAGG.

On either side, add 300 to 400 letters for the homology arms that match the target site DNA.

10 nanomoles of Guide RNAs (ordered online, about $100)

The tracrRNA will be standard. For the crRNA, select appropriate twenty letters to match the homology arm.

Note: Keep the crRNAs and the tracrRNAs separate

10 μL of Cas9 DNA Cutting Enzyme (ordered online, $400 for 20 μL)

P3 Buffer fluid

Xeno-Free Cell Medium

Primary Equipment

PCR Thermal Cycler

Electroporator

The amusing thing about ordering these ingredients online is that the supplier websites offer deals, like "Order 3 and get the 4th free!" just like you're ordering pizza. They have coupon codes. There are almost thirty suppliers you can get them from; Dimitre uses one in particular, but he makes it clear any will do.

Notice we don't need the patient. He can stay at the hospital. Dimitre is not going to inject the patient's body directly with CRISPR. Instead, the physician at the hospital would draw the patient's blood and send it to Dimitre, who will genetically engineer the desired immune cells. Only the cells that are perfectly CRISPRed get injected back into the patient. This avoids all sorts of problems and makes his solution very elegant. For today, we're using blood Dimitre bought at a local blood center. It's analogous to how, on a cooking show, they don't film the slaughter of the chicken.

Surprisingly, Dimitre doesn't even have to sequence the genome of the patient. That's because in the domain he is targeting, there aren't many individual differences. "Only one or two bases vary," he says, "and that little doesn't really matter." In this sense, genetics is a bit like baking–you need precision but not necessarily perfection.

Also like a cooking show: The mixing of the ingredients before it goes in the oven takes about thirty minutes. But like those sneaky recipes from the fancy chefs, sometimes one of those ingredients has to be made *several days in advance,* and the recipe for it is in another section of the cookbook. That's the case here with the blood prep. But the good news is, you can do it up to fourteen days in advance, and you can freeze it, then defrost it when it's ready for use. (Though Dimitre doesn't recommend it.)

There are basically four steps: prep, mix, cook, and rest. It almost sounds easy.

Step 1: Prep the cells.

Strain, enrich, sort, and proliferate the T-regulatory immune cells from the blood. Set aside.

We'll go into the exact steps later, but like a cooking show, Dimitre magically has these already prepped and ready. He's got a little flask that he says has somewhere between 500 million and 1 billion T-reg immune cells in it. That might sound like a lot, but cells are tiny, and so a billion cells wouldn't fill the tip of a pen. To keep them alive, he's using "media," which is basically food juice for the cells. They have to eat just like we do. It's got a key amino acid (L-glutamine) and the transferrin protein. The media is also 5 percent CO_2, like human blood. It's kept at 37°C, which is the temperature of the human body.

Meanwhile, amplify the DNA you ordered in the PCR so you have plenty to work with. A PCR is a kind of genetic Xerox machine, or photocopier, common to labs. It splits the double-stranded DNA into single strands, then rebuilds the missing side—over and over, until you've got millions of copies.

Step 2. Mix the genetic material.

Add equal amounts of crRNA and tracrRNA in a 1:1 ratio. Incubate 10 minutes at 37°C. Add equal amount of Cas9 in a 1:1 ratio with the previous mixture. Incubate 10 minutes. Add 1 microgram of the DNA repair template. Incubate 5 minutes. Pour off media. Wash with saline solution.

I should note the genetic materials are entirely invisible because they are so small. They're suspended in the media. Dimitre uses the smallest pipette and says you pretty much just trust the supplier didn't short you. An eyelash weighs 75 micrograms, so we're describing an extremely tiny amount of material, 1/75 of an eyelash.

Step 3. Cook the genetic material into the cell nucleus.

Spin the T-regulatory cells in centrifuge for 5 minutes at 300 g's. Add the T-regulatory cells to the genetic materials. Mix with 18 microliters of P3 buffer. Transfer into well of electroporation plate. (An electroporation plate is like 12 bullet casings cast together, six by two.) On the LCD panel, key in the EH115 setting. Wait for door to open and load plate. Hit start. In two seconds, remove, return to media.

The electroporator will instantly zap your mixture with an electric field, which momentarily opens up pores in the cells' membrane–allowing the genetic material to enter the cell. The manufacturer claims it also opens the pores to the nuclear membrane, but many are skeptical of this claim, and it's not necessarily needed–once in the cell, the Cas9 has a nuclear localization sequence that gets it into the nucleus anyway. The "cooking" is instantaneous. The pores close again in microseconds.

Step 4. Set aside for 3 to 5 days in incubator.

Now the gene editing happens on its own. You have to give it time, like waiting for a soufflé. Most of the editing happens the first day. The whole time the Cas9 scissors are cutting away and pasting in new genetics, your body is simultaneously trying to get rid of them. Over the course of three days, your body will break up the Cas9 and clear it all out. This is just part of normal protein turnover. It's a good thing, Dimitre explains. You wouldn't want a surgeon to leave his scissors inside. Think of Cas9 as similar to dissolvable stitches. After three days, your soufflé is ready. The T-reg immune cells are supercharged with some new genetic code.

If we really had a patient at the hospital right now waiting for kidney transplant surgery, Dimitre would now purify the cells. The Cas9 will have cut all the cells' genetic code, but only about 20 percent of the time will the new DNA have popped into place. This is actually a really high success rate in these types of cells.

If these are injected into the patient, they're now a "living drug."

Our cells already carry the genetic code to make the natural drug IL10. In immune cells, that gets activated at the site of an immune flare-up. What Dimitre has done (i.e., what he had *me* do) is he's taken that same genetic code and spliced it into a second place in the cell genome. Now that cell is going to make *more* IL10. Far more than twice as much, because Dimitre didn't just drop the code in anywhere. He's lifted the code off a gene that is somewhat active, and copied it over to a gene that's really active. The other thing that will happen is when the T cells reach an immune reaction, they will start multiplying. 100 million cells will proliferate into 100 billion. So the IL10 drug factory will go into massive overdrive, preventing kidney infection.

I'm not allowed to tell you exactly where he put the code, because that turns out to be one of his big secrets. This also helps explain the difference between a good geneticist and a great one. A good geneticist can follow a recipe to perfection. A great geneticist invents new recipes.

In the future, we won't have to take drugs. Instead, we can reengineer the immune cells, then inject them back. Our bodies will make their own drugs. Critically, Dimitre's approach doesn't involve editing your permanent genome. Our bone marrow will always make new immune cells, and the bone marrow's genetic code hasn't been touched in this process, so those new immune cells will be normal.

Since we've got no patient, Dimitre instead injects the cells into a petri dish version of an immune reaction. He shows me how he uses an immunoassay to demonstrate the T-regs are secreting IL10 as designed.

I asked him, if this was done wrong, could it kill someone? I asked it jokingly, but to my surprise he said yes, it conceivably could–if someone put the new code in the wrong place, it could screw up the immune system, and people die from that. But done correctly, it should work almost like magic.

Society has been led to believe that CRISPR can edit any gene in the body. This is a huge overstatement. It's more like, CRISPR can *cut* any gene in the body, anywhere you program it. But whether the new code drops into place is highly variable, depending on what gene you're cutting, how much code you're splicing in, and what loci on that genome you're making the edit.

Not all genes are equally editable. Not all gene edits have the same success rate. So scientists–like Dimitre–have to do a ton of investigation to figure out where on the genome they can splice in code, consistently. The code for IL10 is 1,629 letters–that's no small edit. Dimitre figured out a way to break it into two pieces, and get both pieces to drop in (in the right order). He also had to find a target site that would activate in the right way, and at the right intensity, when an immune reaction was triggered. There's 3 billion choices in a genome.

All of this took a lot of work, obviously. Which is pretty amazing, because–*get this*–Dimitre isn't even doing this procedure for his company. He invented it, then shelved it. Maybe he'll come back to it.

Dimitre's real talent, even beyond this kind of genetic wizardry, is working with T-regulatory cells. Dimitre is what you might call a *T-reg Whisperer*. They are extremely finicky and there's barely any in a blood draw. The conventional T-reg prep kit (available online) only results in a concentration that is 2 percent T-regs. Dimitre took that to 99 percent. Nobody had been able to do that before. He figured this out at UCSF. With deep understanding of their biology, he intuited which antibodies would stain the T-regs so that a cell sorter (which uses magnets) could divert them appropriately. Then he also figured out how to trick the T-regs to multiply 1,000-fold. This was the subject of Dimitre's *Nature* paper. Mastering the full genetics of these T-regs is Dimitre's real goal.

We'll end on the recipe for prepping T-regs from a blood draw.

Prepping T-regs from a Blood Draw

Spices & Oils (all available online)

Antibodies (CD4+, IL2R*a*, IL7R*a*, anti-CD3, anti-CD28)

Cytokines (IL2)

Density Gradient

Xeno-Free Media

Prep Tools (these would be standard in any lab)

Enriching kit

Centrifuge

Cell Sorter

1. Spin the blood in a centrifuge @ 1,200 g's over a density gradient in a 50 mL.
2. Pour out the top portion carefully. These are the PBMCs, including the T cells.
3. Use standard enrichment CD4+ kit per its instructions. Throw in antibodies, add magnetic beads with linkers, put tube in the magnet. Pour out everything else.
4. Chef's tip: going from 2% to 99%. Stain the cells with CD4+, IL2R*a*+, and IL7R*a*. Run through cell sorter.
5. Transfer to T25 flask and stain for FOXP3.
6. Activate the T-regs with anti-CD3 and anti-CD28. Add lots and lots of IL2 to help the T-regs grow and multiply. Can continue for up to 14 days.
7. Empty into T175 flask with media at 5% CO_2. Freeze if necessary.

10

Meet the Mafias Making Buckets of Cash from Illegal Sand New Scientist

How ironic that on a planet rapidly turning into a desert, there is simultaneously a huge shortage of sand, of all things.

We're used to gangs and mafias fighting over control of diamonds, and cocaine, and gun running. Even slave labor or sex workers.

But sand?

All this time, we've been worried about a shortage of oil, then a shortage of lithium, and a shortage of water. More recently we've wondered if we're going to run out of food when the world population surges by 2 or 3 billion people.

But sand?

It turns out desert sand is blown by the wind and so it's too round and powdery to work in concrete. Concrete needs sharp and gritty sand. As the developing world is coming out of poverty, it is constructing buildings and infrastructure like crazy. Roads, dams, bridges. China used more concrete in just the last four years than the United States did in the entire twentieth century.

Some quarries have jaw crushers and cone crushers and impact crushers to make sand from rock, but it takes a lot of energy, making that sand expensive. "Expensive" here is sort of a misnomer–it's still only $25 a ton. But that price has created an arbitrage for gangs and mafias to simply steal beaches overnight and bribe local officials to look the other way while they steal river sand. River sand is apparently optimal for concrete. In India, the sand mafia will drive backhoes up to river shallows and just take away all the sand. One investigation estimated that as many as 75,000 men were being employed to dive down to the bottom of rivers with iron buckets to steal sand by hand.

Concrete is one of the most impactful technologies ever created. Atop the list of the world's most used substances, concrete comes in second only to water.

The essential thing to grasp is that the sand shortage isn't caused by the *population boom.* Futurists so commonly talk about the additional 2 billion people who are going to join us on the planet by 2050–and the strain that's going to put on our capacity limits for food, the atmosphere, the oceans. This fear/challenge is ubiquitous in startup pitch decks. But the population boom–those 2 billion additional people–is actually just the little sister to the much bigger cause of resource scarcity.

Wealth.

It goes unchallenged that it's wonderful for so many people to be emerging out of poverty, by the hundreds of millions around the world. And it's great for the economy, with spillover effects felt everywhere. But it also means a much *faster* collision course with resource limits. Wealthier living uses more. Trees, electricity, water, metals, fish, meat, and whatever we power our cars with. Including–especially–sand.

Right now, there's about 1 billion people living in countries at a "western" standard of living. Earning about $25,000 to $30,000 per capita per year. There's another 3 billion people living in countries tagged with the descriptor "middle income." Most of them are in China, Malaysia, Russia, Brazil, and India. If these countries get to the western standard of living by

2050 (which they are very much on track to do), that alone will increase the demand for global resources by 400 percent. Many organizations study this much closer than I do; all their predictions are reasonably similar.

We call this the "Four Earth Problem." We'll need *four earths' worth* of resources—just to supply the global middle class.

Let's do the math on how the population boom stresses our capacity, compared to the wealth boom.

21% Increase on global resources by 2050 from additional population

400% Increase on global resources from just the BRIC countries catching up to western living.

Now, all this economic activity will be a bonanza for business growth, if it happens. But you can see that the sustainability of future economic growth *isn't just about being nice to the environment.* We've made a big mistake in constructing the sociopolitical argument as a trade-off between the climate or economic growth. We've been led down a decades-long line of reasoning that inevitably characterizes climate-friendly policies as *economic suppressors.* But the desperate need to reinvent the means of production is on every front, with every natural resource. Even sand.

You can imagine a G7 Summit in the year 2040.

GERMANY: Well, Malaysia, it looks like your economic growth is $15 trillion below the number you promised.

MALAYSIA: We implemented robots in every factory, landed a rocket on Saturn, gave everyone a basic income and free tuition, and 95 percent of our citizens now have PhDs in artificial intelligence.

GERMANY: What went wrong?

MALAYSIA: We forgot to reinvent sand.

GERMANY: But every one of your 878 islands is surrounded by sand!!

MALAYSIA: We're down to about 600 islands now. But have you seen our bridges?

It's a common poetic rebuke to recall how people used to live off the land. But as a society, we *never* stopped living off the land. We still live off the land. The land is where nearly *everything* comes from. Every time we fly to see our cousins, and we rent a car and hit a few restaurants and hang out by the pool, we are sucking life's molecular ingredients from the mines and the farmland and the aquifers into the cities–without replacing them.

So the world is facing a philosophical junction it's been at many times before. To the left is a path of austerity and simpler living. To the right is a path of continued and accelerating luxury, while crossing the fingers that someone's going to figure it out. Every time the world has this argument, there's a few wars and a few innovations and someone figures it out. We made it this far.

Earlier in the decade–oh, maybe five or six years ago–a sexy idea got very popular in tech circles. The very, very, *very* shorthand version of it was this: "We don't have to worry we're going to run out, because these smart kids at MIT are starting to figure it all out!" So everyone in tech circles could go back to selling ads and printing money and throwing Social Purpose Parties. Wow, that was easy.

The less shorthand version of it was that we were "dematerializing" our economy. We were still growing rapidly yet using less molecular matter to do it. By making products that lasted longer, such as cars that stayed on the road fifty thousand more miles, we needed new molecules less frequently. By ordering online from warehouses, we were getting rid of physical storefronts, using less molecular matter. And by sharing products (coordinated by our phones), we would fill the world with fewer power drills, boats, bicycles, and wedding dresses. We'd just share them. In the words of MIT's Andrew McAfee, "For the first time in human history, we have decoupled output growth from resource consumption."

It's an absolutely wonderful premise, and it makes people warm in their bellies with good feelings to hear about it. Hug-your-neighbor time.

There's only one little problem with the dematerialization fairy tale. It's not really true.

Inertia is not giving up that easily.

New car sales are steady, and those cars are a third more heavy than they were in the '80s. New boat sales are through the roof. Refrigerators are bigger than ever. Wealth remains the most predictive indicator of resource use. Energy consumption is spiking again. In a lengthy analysis of the usage of sixty-nine materials, the scholars found that only six were being used less, and four of those six were hazardous chemicals that are outlawed or have severe restrictions: asbestos, beryllium, mercury, and thallium. The number one way we've made our economy *lighter* is by layering everything with plastic; even a can of dog food has a plastic layer. But this makes materials far less recyclable.

We indeed are using less phosphate, a key ingredient in fertilizer, but that's also because the phosphate shortage parallels the sand shortage—ships loaded with phosphate coming from North Africa are being taken by pirates—which is driving the price up. The higher the price, the more the farmers deploy it judiciously. They're not using less phosphate because they care about the Four Earth Problem. Thank you, pirates.

Back to the drawing board, perhaps?

So how are we going to *actually* solve it? Maybe a few thousand people can live in yurts, and a few million can ride bikes rather than own cars. But cars and homes are not going out of style. A billion people want them.

I'll just hint here that sand is not the problem. We can make sand from rocks for only $25 a ton. And the earth has plenty of rock.

Almost *everything* we make our physical world out of has problems in the sustainability department. Cement, wood, steel, plastic, gypsum, fabrics, energy, food, films, metals... the list is nearly endless. And then as we blend all those and manufacture them into the material world—the phones, the buildings, the sneakers (22 *billion* pairs of shoes are made every year)... the grocery stores packed with shelves of canned and bagged and frozen organic matter... the vehicles and the furniture—all this *stuff* we love and some we don't—as we fabricate it and polish it and color it and ship it thousands of miles, we greatly increase our sustainability debt.

Remember Yiwu—in China—which Arvind wrote about at the beginning of this book? Yiwu is the pinnacle and the epitome of how the industrialized material world operates today. It's the culmination of 250 years of an industrial revolution. From all around the planet, these raw ingredients are brought to China and processed into usable materials, where they are manufactured into desirable items and packaged into boxes and sent back out all around the world.

The question we're really asking here is this: What's the Yiwu of the future? Will we still harvest and mine these legions of materials? Will the machines of the future still be made of metal, with gears and presses and robot arms? Will the factories of the future still have kilns that reach three thousand degrees Fahrenheit? Will we still use hazardous chemicals to transform matter?

Every year, in the United States, we make and use a *billion* pounds of hydrogen cyanide. Hydrogen cyanide is precisely what you think it is–it's a form of cyanide. It's a chemical warfare agent. It's banned by the Geneva Protocol. But despite that, we make and use a *billion* pounds in industry. We make it in blast furnaces at about 2,200°F. We use it for all sorts of things–hundreds of ways. We use it to finish metals, to make synthetic fibers, to make plastics, fertilizers, and to make other chemicals. The industrial revolution of machines has long had a silent partner, which is the chemical revolution.

In the 1985 movie *Back to the Future*, the final scene has Doc needing to fuel his DeLorean. He digs in the trash, grabs a banana peel, a few unseen bits, and a half-empty can of Miller Lite beer. Doc dumps these into his car engine (including the beer can) and *voilà!* Marty and Doc fly away. It's a funny scene, but one we've rewatched quite a few times around the office, because it hints at the notion if you strip your trash down to its fundamental atomic and chemical elements, they can be reused any way you want.

Though the industrialized world can seem very complex, all these things are made out of a fairly small tool kit of fundamental elements. 97 percent of our bodies are just oxygen, carbon, hydrogen, and nitrogen. Add in a short list of other items–calcium, phosphorus, magnesium, sodium, potassium, chlorine, and sulfur–and you're well over 99 percent. Trace amounts of metals round it out. Just twenty-five elements make up all living things. The material world around us is more diverse, but really not *that* diverse. Just four elements make up 90 percent of Earth.

Transforming matter at the elemental level is exactly what chemistry does. But it's also what *biology* does. In most people's minds, biology and chemistry are two different fields of science. But in fact they are both manifestations of molecular physics, and it's useful to think of biology as a kind of elegant chemistry. When I say biology, don't think of human bodies or plants–think of bacteria. There's a trillion species of bacteria, and they can turn just about anything into just about anything else.

Bacteria don't just make yogurt. They can bind carbon to silicon. They can make electricity. They can absorb gamma radiation. They can eat minerals and excrete acids, or eat acids and secrete minerals. They can emit light. They can make superglue. They can eat electrons and breathe out metals. They can align themselves like a compass to Earth's magnetic field. They can process gold ores into tiny gold nuggets. They can turn sunlight into bioplastic. There's even a bacteria that eats rock and poops out sand. Nice, coarse, gritty sand.

For almost any chemistry experiment you can imagine, there's a species of bacteria that does it naturally. Bacteria are a kind of algorithm of the material world. Formulas. $2x + 3y = 5z$.

This trillion-species tool kit is now under our control. In every species of bacteria, its DNA is like its computer code, telling it what to do. That genetic code can be reprogrammed and altered any number of ways. Most bacteria eat sugar, but they can be reprogrammed to eat CO_2. Or methane. Or eat the plastic out of the ocean.

Now, you may be wondering, can bacteria–can *biology*–produce things fast enough? They're so tiny. Today's industrial revolution features superfast efficiency. Amazon ships 1.6 million packages every day. Across all their factories, Apple makes ten iPhones *per second*. We cut down 15 billion trees a year, and we pull a trillion fish out of the oceans. It's all about speed and volume. Isn't biology kinda *slow*?

Actually, no. In every one of your cells, there's between 10,000 and 10 million ribosomes making proteins. Transcription is 40 to 80 nucleotides per second. Proteins are built at 20 amino acids per second. Each cell makes 100 sextillion proteins a day. That's 10 to the 23rd power. *And that's just one cell!* Multiply that by the 37 trillion cells in your body.

Each one of these proteins is essentially a tiny robot, often extremely sophisticated. Take the ATM protein, coded by the ATM gene. It constantly flies along our chromosomes, riding them like a zip line, searching for DNA damage. If it finds damaged DNA, it sends out signals to stop the production line. The ATM protein is light-years ahead of the security robots we use to guard factories. And every ribosome in our body can make it–in about ten seconds.

Bacteria are even faster, and the magic of bacteria is that–while each one is a tiny factory–they simultaneously split in two and divide, constantly. So we can create factories of any size we want. The slowest bacteria double in half a day. The fastest double every ten minutes. The most common industrial bacteria, *E. coli*, doubles in volume every twenty minutes. We tend to think of microbes as being invisibly microscopic, and they are–but in a few days they can fill tanks the size of houses, producing whatever we want them to produce.

So, we don't really know what the Yiwu of the future is. It's something we're exploring, experimenting with. We have a company that does pretty much what Doc's DeLorean of the future did–they use electron-breathing bacteria to turn ordinary restaurant food waste into hydrogen fuel cells for hydrogen vehicles. We have a company that replaces animal leather–with a remarkably similar bioleather made from fungi. We have a company that could eliminate the need to cut down trees to make wood; they make wood by combining flax fibers with bioresins. It's stronger than steel, moldable like plastic, and lighter than carbon fiber. Yet another grows airborne bacteria into fishmeal, for aquaculture–so we can stop pulling all the fish out of the oceans.

A company called bioMASON (not one of ours) grows construction bricks from sand and bacteria–which is also how coral reefs build themselves. These biobricks grow *way* faster than coral, though; they're finished in four days, which is two days faster than firing bricks in a kiln. Imagine one day being able to grow an entire building in place, rather than assemble it.

All this might sound wild and crazy. But it really comes down to understanding that we are *not* actually turning one thing into another thing. All these things are made of the same littler things–the same fundamental elements. It's something we all did as kids, taking apart a Lego house and using the parts to make a Lego car.

When the Industrial Revolution started in Britain 250 years ago with steam-driven yarn spinning, it made the country fabulously wealthy. Due to fear that their technology would get copied, it was illegal to export their machines, and for almost fifty years, Britain had a remarkable head start on everyone else. Today, with far more sophisticated machines, China is now growing fabulously wealthy by perfecting the process. Factory automation will continue to power this wealth creation for some time. But we still can't help but wonder–what will be the industrial revolution that follows? We're not going to dematerialize the world, we're going to rematerialize it.

11

What's Your Purpose? Finding a Sense of Meaning in Life Is Linked to Health NPR

Epidemiology doesn't get much love these days.

Epidemiologists study people's health over the long term and try to sort out why they're falling ill.

But in an era of CRISPR babies, longevity drugs, and bioprinting replaceable organs, epidemiology isn't exciting. You just don't read as much science news anymore where volunteers were studied for a long time, in their natural habitats, as a way to learn something radically new about health. It feels old-school. An epidemiologist's equivalent of mass spectrometry is a survey questionnaire. Write your name, check some boxes, come back years later, do it again. *Is that really going to tell us anything?*

Three epidemiologists from the University of Michigan got their revenge with this one. They studied almost seven thousand people over the age of fifty for a half decade. They asked a really unusual question of these people.

It didn't matter if you were rich or poor, black or white, healthy or ill. Across every cohort, controlling for all variables, if you had "a sense of purpose" you stayed healthy, and if you didn't have a sense of purpose, you didn't stay healthy. It was more powerful for your health than exercising every day. It was more important than smoking or drinking. It was pretty much the *strongest determinator of your future health.*

How did they measure if people had "a sense of purpose"? They basically just asked them, yes or no–"I have a sense of direction and purpose in my life." On a scale of 1 to 6. Seven questions, all pretty much the same. "My daily activities often seem trivial and unimportant to me," or "I don't have a good sense of what it is I'm trying to accomplish in life."

No electron microscope of "purposeful" cells. No X-ray of the soul. No genetic marker for inspired motivation. Nope. *Just checkboxes.*

And yet. The finding was a stunner. And it was inarguable, because they'd even measured the possibility that having chronic illness could cause you to lose touch with your purpose–which it does, somewhat–and subtracted that reverse effect. Also, this study wasn't an outlier; there had been nine previous studies leading up to this one. The epidemiologists published it in *JAMA*, which is one of the most credible medical journals there is. It's the Tom Hanks of science.

Nobody was arguing if you've got "a sense of purpose" you can survive getting hit by a bus. Or that you can start living like Bill Murray in *Groundhog Day*, throwing yourself into dangerous situations. Just that of *all* the common things we do to take care of our health, especially as we age–improving our diet, exercising regularly, making sure we sleep well, cutting out smoking and alcohol–none of those matters as much as feeling like there's still a reason for you here on Earth.

You cannot measure this sense of purpose in the body. It's not a biological thing–not that we know of. It's likely a perception of the brain, a meta-awareness of the general pattern of our actions and the role we play in the environment around us. The key word is "sense," meaning cognizance. A fork has a purpose, but not a sense of its purpose.

But somehow, we are left to conclude, this master perception trickles down from its neuroelectrical state in the brain through disparate biological mechanisms, inciting action chains that defend us from decline–moderating stress hormones, secreting cytokines to soothe the immune system, or fending off free radical DNA damage. Mind over matter.

You could go through the whole haystack of the body, looking for this needle of purpose, and never find it. Yet it's somehow there.

It's important to note that the volunteers in the study didn't have to put words to their purpose. There was no blank line or empty box that said, "If you answered 'yes' to having a sense of purpose, tell us what it is here." And there was no Michigan psychology student reading the questionnaires, rating the various purposes for integrity and saying, "Well, that's a shitty purpose, it doesn't count."

It would be fun to go to the annual *JAMA* awards dinner and see the epidemiologists with their checkboxes roll up to the bar with all the neurosurgeons and geneticists and microbiologists. *Yeah, we discovered the secret to staying young and not dying.*

Wow! How did you find it?

A checkbox.

Over the course of this book, you're going to hear a lot about new ways to protect people from disease and rescue them from death's door. Some of them are going to seem magical. We thought it worth including this point as a perspective; it's important to appreciate that some of the magic is still well beyond their reach.

In 1976 there was a famous study done by Ellen Langer in a nursing home. They randomly gave half the residents a plant to take care of. The other half didn't get a plant. Eighteen months later, the plant caretakers were dying at half the rate. Further, the group with the plants were happier, more energetic, and healthier. It strikes me as a great illustration of the power of purpose.

We live at a time where the concept of "purpose" has been hijacked. Bludgeoned into nothingness through imprecise overuse. Every millennial coming out of school is looking for a job that offers "a sense of purpose."

And almost every corporation has learned to use social media to highlight their purpose-driven marketing campaigns; they donate military care packages, ship supplies into hurricane relief zones, and offer pro bono services to nonprofits in a kind of symbiotic gestalt of doing good. There's a whole cottage industry of experts who coach corporations to chill, stop being defensive, and open their arms to a little social capitalism.

In this way, purpose has been industrialized. Mass manufactured. Nobody is in sales anymore; they're in customer relationship management. Nobody is in marketing, they're in customer engagement. Skittles has 22 million followers.

There's this old parable about the Three Bricklayers. They're laying bricks all morning, and when they finally get a break, one guy asks the other two, "Why are you doing this job?" The first guy says, "I'm doing it for the wages." The second guy says, "I'm doing it for my wife and kids." The third guy looks up at what they've been constructing all morning, which is a church, a house of worship–a place to get in touch with one's highest self–and says, "I'm helping to build a cathedral."

Now, when most people hear this parable, they think the third guy has the right answer, and the first two guys have the wrong answer. That's the simplistic lesson that most people jump to, led there by their mythic notions of calling. But that is not the *real* lesson of the parable. In fact, all three men have a sense of purpose–have a "cathedral," if you will. The third guy has the Cathedral of Spirituality. Good for him. But the second guy has his, too: the Cathedral of Family. And the third guy has the Cathedral of Self-Sufficiency. Those are all good purposes. Those are all *right* answers.

The real lesson of the parable is, notice what no man answered. Not one of the three said, "I just love laying bricks."

Purpose is *constructed.* It's a conscious mental abstraction. It's not intrinsic. It's deeply personal. It doesn't just show up on your doorstep. You have to lay some bricks to have a purpose.

Here's the seven checkboxes on the epidemiological study.

2012: 7 ITEMS (Q35a-Q35g)

(Please read the statements below and decide the extent to which each statement describes you.)

Q35a I enjoy making plans for the future and working to make them a reality.

Q35b My daily activities often seem trivial and unimportant to me.

Q35c I am an active person in carrying out the plans I set for myself.

Q35d I don't have a good sense of what it is I'm trying to accomplish in life.

Q35e I sometimes feel as if I've done all there is to do in life.

Q35f I live life one day at a time and don't really think about the future.

Q35g I have a sense of direction and purpose in my life.

Coding:
1 = Strongly Disagree
2 = Somewhat Disagree
3 = Slightly Disagree
4 = Slightly Agree
5 = Somewhat Agree
6 = Strongly Agree

Notice the key words in these life-extending questions: The future. Plans. Working. Active. Direction. Accomplish.

Okay, I'm going to call a premature end to this chapter. It's a failed experiment. I tried writing the rest of the chapter multiple times, but no matter what, I kept boiling the ocean. I lost focus and became too abstract and tried to wrestle with the entire future. Sometimes, with an experiment, you just need to acknowledge it isn't working.

11:20 PM
99%
Messages
Arvind
Details
I'm having such a hard time writing Wittenberg
I keep boiling the ocean
hahahah
where are you?
in this beautiful town in eastern Germany. Zittau.
this is the first chapter you're writing without actual science in it...
Yes. And I feel unrooted without it.
but the first principle of science is there. the very thing that gave rise to science. Questioning.
that's it
Text Message

12

Meet the Pope's Astronomer, Who Says He'd Baptize an Alien If Given the Chance

National Catholic Reporter

I'm standing outside All Saints' Church in Wittenberg, Germany, looking at its famous doors. It's a fairly humble building, tall like all Late Gothic architecture of the era, but very much a church, not a cathedral. It was here, 502 years ago, that a local professor of theology unintentionally invented zines, and accidentally kicked off a massive revolution against Catholicism. The church will open in a few minutes, and what I'm struck by, more than anything, is how few people are here. It's just me and a young Chinese family.

I wasn't expecting Notre Dame crowds, but I figured to spot at least a Lutheran Pilgrimage tour bus or two, stopping along their route between Berlin and Frankfurt. Alas, *nobody* is here but us.

The main entrance is now on the other side of the building; the doors that Martin Luther nailed his *95 Theses* to are protected by a short iron fence.

Actually, those original doors have been replaced, by darkened bronze doors that have the *95 Theses* cast in Latin into the panels.

I had planned to say that I'm not here for the same reason everyone else is. But with nobody else here, I'll put it this way. I'm *not* here because it's the birthplace of Protestantism. I'm here to test this hunch I have–that Wittenberg is the birthplace of the *scientific revolution*. That modern science traces back to the very spot on Earth I'm standing on today.

It's a theory I've wanted to think through. And the best place to do that thinking is Wittenberg itself.

The rough sketch of my idea goes like this: For science to take root and spread, it needed to be socially permissible to question the Church's authority and refute its official explanation of the nature of life. When Martin Luther triggered a civil war with the Catholic Church, he kicked off this century of questioning. This turmoil of social change loosened the psychological firmament, making it ripe for exploration of all sorts, including the most radical idea of all–that the nature of reality could be determined *only* through empirical proof.

In other words, the scientific revolution could *not* have happened unless there was a social revolution that preceded it.

When the church opens, I take a seat in a back row pew. A few others stroll in now. I don't know why I find it funny that there's a gift shop, but I do. Maybe it's the Martin Luther Lego minifig they sell.

The previous night, in Dresden, I had dinner with a Lutheran pastor to explore my theory. I learned this was not Martin Luther's church; he didn't preach here. He lectured at the City Church nearby. He had started out as a law student at the local university, but he had always been unsettled, he actively worried about his own death, and he had a deep craving for grounding truth. Catholicism's threat of eternal damnation scared the shit out of him. He was a literalist. One night, during a lightning storm, it wasn't that he saw God as much as the lightning bolts persuaded him he was going to hell if he didn't become a monk and try to live the purest life he possibly could.

This happens to a lot of us–we have an epiphany, make a grand resolution, and then the next morning we wiggle our way out of it and go back to our normal life. But Luther's sense of obligation and duty was strict. He felt he couldn't back out of his vow. When he started at the monastery, this kind of pious extremism defined him. He fasted more than others fasted, he prayed longer than others, and he confessed constantly, often to glancing thoughts that had momentarily crossed his mind–stuff he was constantly told, *You don't need to confess that. That's not a sin.*

At the monastery, all monks were given a Bible, and memorizing that thing was super important, because they took it away at the end of the year. It wasn't just that the Bible was in Latin rather than German that restricted access to God. Nobody was allowed access to Bibles for occasional reference. Not even monks. They could only study interpretations written by theologians like St. Augustine. Luther's memorization skills were legendary, but when his monastery took away his Bible it crushed him. If it had said in the Bible, "Burn this after reading," he would have been fine with it. But not one line in the Bible said anything about how important it was to go through life *without* a Bible.

Luther had a tendency to see hypocrisy and mendacity everywhere he looked. He got a mentor, and this dude volunteered him for a trip to Rome on some monastery business. He thought it would clear Luther's head. Luther piously walked every step of the nine-hundred-mile journey, and was deeply disheartened by the incredible opulence of the way people lived in the Vatican, and the tales of immorality he heard. Nobody was good enough for his confessions. He began to question the very idea that pilgrimages and sacred rituals could get a soul out of purgatory.

I leave All Saints' Church and walk toward Luther's monastery. Wittenberg is an austere place. It's very isolated. One of the nice things about Germany is that the roads between cities are not through endless suburbs and edge cities. Towns actually stop and then you drive through fields and forests. But getting to Wittenberg, this was especially the case. And I realized this isolation was probably a critical ingredient to the Reformation. Luther could think for himself here, and the people of Wittenberg weren't directly under the foot of authorities; their bishop was forty miles away, and the archbishop was all the way down in Frankfurt.

The archbishop had pretty much paid for his position, but he also wanted to be named a cardinal, which was going to cost him millions more. Pope Leo was raising money to rebuild St. Peter's Basilica. So the archbishop borrowed the money from a banking family, and much like the industry lobbyists of today, the archbishop got permission from the pope to sell free passes to heaven to pay back his debts. He could broker deals with all the grain merchants, copper mine owners, and shipping industrials to clear their record of sins for a major donation.

Into the town of Wittenberg came a monk, operating on behalf of the archbishop to cut deals. Smaller indulgences were already common, but the money raised was always spent locally–a bridge had been built, and the university improved. But funding Rome ticked everyone off. Parishioners complained to Luther, and you can imagine how reprehensible Luther must have found it. When he posted his *95 Theses*, it was a common practice to post notices on public doors to stir debate. Back then, church doors were the equivalent of today's Twitter and news crawls. Reading the theses, it's clear ninety-five was overkill. He easily could have made his points in under thirty. His main points could have been made in a mere two theses. But Luther had been a lawyer, and his legal mind is evident in the document, as he foresees every possible argument and wiggle room, and tries to rebut those a priori.

His *95 Theses* would get him condemned by the pope and eventually excommunicated. But his friends immediately printed leaflets of the *95 Theses* in Latin, and a couple months later, in German, when it went viral. Within two weeks, they were all over Germany, and within two months, reprinted all over Europe. Luther was on the run, even as he tried to help prevent his religion splintering outright. He translated the Bible and published it in German. Luther even freed some nuns from a monastery, smuggling them out in herring barrels, and he married one of them, beginning the practice of clerical marriage. He and his wife moved back into the monastery where he had studied before, the exact place I've reached on my twenty-minute walk from the church. There's a few more people here, but this is no Western Wall.

The exhibit inside that touched me the most was unexpected, and something the Lutheran pastor hadn't mentioned. It was a Common Chest, a big dark treasure chest with pretty elaborate locking mechanisms on three sides of its lid. It sat not in All Saints' Church, but in City Church. Luther had abolished begging, both by the poor and by traveling monks who'd taken vows of poverty. Instead these people were taken care of by the contributions to the Common Chest. From this chest came the money for the orphanage, the hospital, the schools, and support for the poor.

Over the next decade, Lutheranism would take off in the north, Anglicanism in England, and Calvinism down in Switzerland. Western Europe plunged into chaos. And this is where it dovetails into the scientific revolution. Historians normally begin the scientific revolution with Copernicus, the Polish mathematician and astronomer. By 1517, Copernicus had a notion that the sun was the center of our solar system, but he hadn't worked it out in any detail. He'd written a one-pager that he let only a few astronomer friends see, to explain why they should start tracking eclipses. He was more focused on a theory of money supply.

At the time, the Church *was* science. It declared an explanation for all that needed to be explained. The Church ran the education system and controlled what people learned. It wasn't technically in the Bible that Earth was the center of the universe, but there's a reference in Joshua 10:11–13 that kinda hints at it. That's certainly what was taught, and that's what was accepted, even by Luther. Ironically, Luther had even heard about Copernicus's theory, and in his notes he made fun of it. So certainly Martin Luther didn't directly encourage the astronomer.

Nevertheless, Copernicus wasn't far away from Luther, geographically, and Lutheranism began to spread through Poland. The militaristic Teutonic Order adopted Lutheranism in Poland, partly as a power move. Copernicus didn't become a Lutheran, but many of his friends did. And I have to think that this social upheaval made *more* change possible. I have to think this change in society's belief system made Copernicus feel a little less heretical for developing his radical theory in depth. His model threatened the Church's official explanation, but at least questioning the Church was in the air. Copernicus didn't generalize his theory into a movement, declaring that empiricism was the source of truth–that came later, from the minds of others–but he and his colleagues did try to verify his theory empirically. And when Copernicus's book was finally published, it was the Lutherans who would embrace it.

The Protestant Reformation and the Scientific Revolution are intertwined much like all chicken-and-egg problems. One couldn't have happened without the other. Social change sometimes triggers technological change; sometimes it's the other way around. But they always go hand in hand. No technological revolution fails to be directly connected to a social revolution, either just before or just after. Science always needs intellectual freedom to flourish, and every scientific breakthrough has to fundamentally question the stranglehold of scientific consensus. The freer people feel to question, the faster science will advance.

Today, the technological changes afoot may seem mysterious, but they're far less murky than the social changes that will accompany them. Mass migrations, autocratic crackdowns, billion-strong protests, financial restrictions, new taxes, and water wars may be the *least* of it. It's great that the pope now has an astronomer, and Pope Francis has even spoken publicly about artificially intelligent robotics being deployed solely in service of humanity. Because we're *for sure* going to need religion's powerful voice to hold society together.

What makes the social changes even harder to see coming is that they can happen almost overnight, and the littlest of things can unleash them. Technological change works a slow, fifteen-year road that begins in academic labs, shifts into commercialization, and eventually gets globalized. It took Copernicus two decades to work out his astronomy. It took Luther twenty minutes to walk down to the church and post his theses.

13

There Is an Absolutely Gigantic Rogue Planet Wandering Our Galactic Neighborhood Science Alert

It's dusk and I'm standing in the middle of the California high desert, sidestepping bitterbrush, staring at a massive scatter field of hundreds of small aluminum radio antennas, each the shape of a little teepee frame floating a couple feet above the desert grit. The antennas belong to Caltech, and were designed by a professor and his graduate students, who've now left for the night. I'm not sure if the antennas can sense my presence; they've got other things to do. They're looking for life on other planets, and I'm looking at them, thinking about what I'm really seeing and witnessing. Tonight is a full moon; its brightness will obscure any good look at the Milky Way. I snap a lot of photos anyway, mostly of the antennas, which glow even whiter in the photos than they do to the naked eye. A jackrabbit crosses the antenna field.

I'm *not* seeing life on other planets. (That's what the computers in the nearby shipping container do.) I'm a human; I've got only these two eyes to gander with. I'm witnessing the silent *search* for life. I'm witnessing mankind's restless nature.

And more abstractly, I'm seeing an insane feat of human imagination.

Because it takes incredible imagination to think a tiny blip in radio waves hitting Earth from other solar systems could be important, a signature of exoplanet habitability. And it takes even more imagination to build a planet-searching machine for less than a million dollars, with just some wire, aluminum sticks, and computers.

Like everyone my age, I watched the first lunar landing on my parents' black-and-white television. I drank Tang and ate Space Food Sticks at school. I played "moon colony" in our backyard with the neighborhood kids. But it didn't carry with me to college. I didn't think the space shuttle really was all *that* cool–certainly not compared to lunar missions. SpaceX didn't get my heart racing. I mean, I thought it was great that a private company jolted our government out of its doldrums. I felt that aspect was heroic–but I wished our government took the same interest in exploring our ocean depths.

And as a journalist, I've had my fair share of assignments on space-related stories. But I'm no space fanboy.

I have lots of friends who go crazy for space stuff. They are four-exclamation-point people. Space!!!! They meet an astronaut and their knees go weak and they can't think. I'm naturally more of a three-dot-ellipses person. Space...

So, as an experiment, I decided to see if I could catch the space bug. Just to see if I could be converted.

And to do this, I went searching for life on other planets.

It turns out, you don't need to be a billionaire to look for life "out there." Instead, you just need to make friends with Gregg Hallinan. He's a professor of astronomy at Caltech. With about $70 worth of unleaded gas, from anywhere in the Golden State you can drive to his crisp white shipping container in the Owens Valley and take a gander outside our solar system. If you want to look beyond the Milky Way, it's now about as easy as changing the channel. "Here, from 2.5 billion light-years away, we've got a performance by a couple of black holes." Last year, Hallinan's team discovered the rogue planet, and now they were looking for more.

The Owens Valley is high desert, a xerophytic ecosystem fed by bajadas and alluvial fans that sprout next to nothing–bitterbrush, burrobush, buckwheat, creosote, and the occasional lizard or jackrabbit. It's nestled between the Sierra Nevadas and the Inyo Mountains, where the legendary 4,500 bristlecone pines grow. The unforgiving setting seemed fitting. I showed up in the Owens Valley thinking that finding life on other planets was aspirational, lonely, impossibly poetic but implicitly futile. Worst of all, *terracentric*, which is to an astronomer what *chauvinistic* is to a feminist studies scholar.

Damn, was I wrong. Fundamentally, Hallinan and his graduate students have taken the solitary search for habitable planets and *automated* it. They'd made a planet-searching machine.

Hallinan's telescope truly defies expectations and defies the trend in astronomy, which has always been toward bigger telescopes that cost billions of dollars, while moving higher up the electromagnetic spectrum to really energetic wavelengths–gamma rays, X-rays, gravitational wave events. Hallinan is instead using the lowest-energy end of the spectrum, radio waves. Even lower on the dial than FM radio. And rather than one big telescope, he and his team stuck 288 antennas in the desert floor, spread out over hundreds of yards in each direction. These antennas are wired together with eighty-eight kilometers of cable, which run into the white shipping container. Inside that container, it's frigid. The air conditioner on top is so big it screams like a jet engine. The cold is to protect the supercomputers that make sense of all the radio waves hitting the antennas, 200 million times a second.

This antenna field monitors the entire sky for auroras from other planets. Much like the Aurora Borealis–but in this case it would be Aurora Teegarden b, or Aurora Kepler-47c. If you were on those planets, you would see northern lights; what Hallinan's antenna hears is just a blip. But it's a telling blip, because auroras are caused by coronal mass ejections sparking into a planet's atmosphere. Atmospheres are essential for protecting life on other planets because they block highly energetic solar radiation that would otherwise tear DNA apart. If Hallinan's array picks up an aurora, they'll know life could be there.

In fact, the planet we talk most about visiting, Mars, is a cautionary tale. Mars very likely had life on it several billion years ago. But it lost its magnetic field when its molten core died out. Gravity plummeted, and the solar wind blew away its atmosphere. Hammered by gamma rays, life on the surface of Mars disappeared from sight. So even if we create a colony on Mars, we'd have to live under a lead shield, or be torn apart by radiation.

Regularly, the sun fires off coronal mass ejections, a billion tons of plasma hurtled our way at more than a million miles per hour, Hallinan describes.

There's an important difference between Hallinan's aurora scanner and going to Finland hoping to spot the northern lights. He's removed any need to be in the right place at the right time. The scanner monitors all the 4,011 known planets within seventy-five light-years–and as that number climbs to 20,000 in the next couple years, it will monitor them, too. If any of them spark with an aurora, he'll know.

My visit happened to coincide with a big project review meeting, where a board of astronomers was evaluating Hallinan's progress and budget. Because the board was there, so too were most of Hallinan's research students, even a few undergrads. Unexpectedly, I got to hang out with about twenty astrophysicists. They were of every personality and age, from eighteen to seventy-five. The one thing that was clear was how much they *love* space. They talked about it like parents who brag a lot about their children's accomplishments.

"The forty-meter scope could pick up a cell phone on Pluto."

"A teaspoon of magnetar weighs a billion pounds. It's pure neutrons."

"Space provides natural experiments–with one hundred times more energy than the CERN supercollider."

They were all humming with excitement because just a couple weeks prior, one of their professors had discovered a galaxy 7.9 billion light-years away. Major discoveries could happen at any moment.

Of course, I didn't want to talk blazars and quasars. I wanted to talk about *life*. And while none of them were biologists, they were game for the conversation. It always felt like astronomers were holding out the carrot of "maybe there's life out there" as a ruse, just to keep us interested in their lifeless rocks and invisible gamma rays. I'd been taught that life on Earth was an incredibly fortuitous accident, a once-in-a-gajillion-years miracle. Before that moment, 3.5 billion years ago, Earth was inorganic. After that moment, things came alive. Strange chemistry was involved, likely deep in an ocean vent.

And while that may be all true, here's what I didn't know: Stars don't just shoot out energy and light. When they're young, they shoot out amino acids and nucleotides–the raw material of DNA, RNA, and proteins. These are carried by the solar wind throughout the solar system. The building blocks of life *come* from suns. Earth wasn't just extremely lucky to have these incredibly elegant Legos here. Early in the formation of a solar system, all the planets get showered with them, especially the rocky planets near the sun. Earth isn't special.

Comets carry these amino acids between solar systems, and meteorites protect them from gamma rays. We've found RNA building blocks in solar systems four hundred light-years away. Water is nearly everywhere in the universe, as is carbon dioxide and other common gases. Oceans on Saturn's moons are brewing with nutrients. And those incredibly lucky extreme conditions that science believes gave rise to unicellular organisms on Earth are an ordinary gaseous explosion in space. Bacteria can spore and survive at least thousands, if not millions, of years. Even photosynthetic bacteria has survived space travel.

Looked at this way, it becomes highly improbable that Earth is the only planet with living biology on it.

Another way to characterize it is that our chances of finding a habitable planet are roughly equal to the chances that Christopher Columbus, sailing east, would accidentally discover the Americas. It's not a question of *if.* It's only a question of *when.*

This doesn't mean life on that planet would be instantly recognizable, even under a microscope. Alien DNA could be built with alternative nucleotides, which form with equal amounts of energy so are just as likely to be present. The DNA backbone could just as easily be a Morpholino, or PNA, a peptide nucleic acid, rather than our sugar-phosphate frame. And they could use different amino acids as building blocks. But these variables are things that Earthling geneticists already play with, in the lab, synthetically making them all. If we needed to design a new life-form to match another planet's, with eight nucleotides and noncanonical amino acids, we could do it. But none of these things are technically what we mean by *life*–they're just the programming code for life. And just like people learn to speak languages in foreign lands, or computer programmers employ different instruction languages, so too could biologists today program life for other planets–if we ever meet it.

Life, though, really means a cell membrane. In biology, what defines life is the ability to self-reproduce. And only things with cell membranes can reproduce on their own. Since it took our planet a couple billion years before life began–before unicellular organisms appeared–we can narrow our search among the planets by looking for solar systems that are at least a few billion years old.

Probably the coolest thing I learned on my visit was that Gregg Hallinan's aurora scanner is sort of just a working prototype; NASA is hoping to clone it on the far side of the moon sometime this decade. By looking at outer space from the moon, they'll be able to avoid all the radio interference of Earth's atmosphere and will receive much stronger signals.

After taking more photos of the antennas, I wander the dustland around Caltech's compound. There's a lot of other radio telescopes here–more conventional ones, of every size, some of which are still in use and some of which are decommissioned. They all have that NASA early-space-era design aesthetic, gleaming white. The biggest one is 130 feet tall and wide; it's been here fifty years, and it still checks in on 1,800 blazars twice a week. It's sort of the celebrity here, and it's even been in Hollywood movies. It's quite beautiful against the darkening sky.

As I watch the full moon rise over the Inyo Mountains, I realize how my day here has moved me. It's not that I fell in love with space during my adventure to the desert. Rather, I just didn't know space could be so full of life. And the nature of life has been something that's always interested me.

Earth is 4.5 billion years old. Life has had booms and busts for a solid 3.5 billion years, with at least six mass extinctions. But we humans, we just got here, near the end of the party. We've got maybe half a billion years before the sun starts dying on us. A billion years at most before the oceans are gone and the surface is inhospitable. It's both a nearly infinite amount of time, in human years–think of all the history that will be written!–and yet also an incredibly short time, in planetary lifetimes. Like walking in late to a movie. What, it's over? We just got here.

When I get home I call several people working with NASA. "I just had dinner with NASA administrator Jim Bridenstine," says University of Colorado Boulder astronomer Jack Burns. Burns was on the presidential transition team for NASA and constructed the space program for the incoming Trump administration. Burns has been envisioning a radio telescope on the moon since the 1980s. "But until Gregg built his system in the desert, we just didn't have the killer app that would justify it."

And if we find another planet like ours?

"Well, we can't get to it," says Burns. "We can't even get a probe there. The chemical propulsion rockets we have now are fine to get around the solar system. But we would need leaps in propulsion technology."

His words turn speculative, but he inevitably appeals to our restless nature. "It could happen by the end of the century. There's a lot of ideas. People are working on them. There's no limit to the imagination."

14

Just a Few Sips of Soda or Juice Daily May Up Cancer Risk by 18%, Study Finds USA Today

In 2014 I was at a conference held by DARPA, listening to a Navy colonel speak about their recent efforts to connect the internet with your mind, when my phone rang. It was my sister. I stepped out of the conference hall and answered, "Hi, Tiller!" (my childhood name for her, derived from Atilla the Hun).

She was crying. "Mom has cancer," she said, between sobs. "They found cancer in her uterus. She went to the doctor because she was bleeding after menopause and that's when they found it."

My sister is an eye surgeon specializing in the cornea. We could have had a technical conversation, but we did not. We were scared and heartbroken at the thought of losing Mom.

Many families struggle with why a loved one gets cancer, especially when that loved one didn't lead a lifestyle that raised her risk factors. Mom didn't spend her life around chemicals, or dyes, or filling her lungs with particulate matter. She didn't sit at a desk all day (which increases the risk of uterine and colon cancer). The desire to attribute a *cause* to the effect, cancer, can be so strong that many families are plagued with guilt, second-guessing their mentality, their lifestyle, the things they let stress them out. The randomness of cancer is often very hard to accept.

And even though I knew the fundamental pathways behind this randomness, I too struggled with *why*.

How it happened is different from *why*. A single cell in my mom's uterus had a mutation occur in its DNA that escaped her DNA repair mechanisms, which started a cascade of molecular interactions that caused it to never die and begin dividing. Uncontrollably. This first cell became two, then four, then eight, then sixteen, and so on, all genetically identical. Quickly the mass of cells began to pick up more mutations that help them survive. Through all of this, the opening first stage of cancer, my mom didn't feel a thing. Until it began to invade the uterine lining, which caused her to bleed.

The hysterectomy was scheduled a week later. It went smoothly and we all breathed a sigh of relief.

Cancer isn't caused by fruit juice. Or alcohol. Or sunlight. Or any of the things you hear in the news. Cancer is caused by mutations. Mutations that pile up until runaway cell division occurs. We have random DNA damage happening in our bodies all the time. A single strand of your DNA breaks over 55,000 times per cell, *per day*. This is a staggering 2.03×10 to the eighth, or 2 *quintillion* times in your body per day. When a cell divides, about once a day, the odds of making mistakes in making a copy of the DNA for the new cell is one in 10 million. But we have 3 billion base pairs that the cell needs to copy, leading to thirty thousand errors per cell cycle. With 37 trillion cells in the human body, that adds up to an astounding amount of errors.

On top of that base rate of mutation, our lifestyle and environmental factors cause genes to get silenced–not across the whole body, but cell by cell. The chemical composition of a genetic nucleotide gets altered a little. This makes it more likely to get a mutation that is then passed on in cell division.

The way the media describes our genetic code, it routinely conveys the old idea that our code is consistent across all cells in our body. When you send your cheek swab to a service to have its genome sequenced, the genome they show you in the reports *is just the average*. It varies some. And you gave them samples only from your cheek, not the rest of your body.

The basics we are taught in biology class usually fail to capture this essential truth: Mutations are constant. They are *both* random and not. They occur at a predictable rate, and the odds of one mutation is known, sort of like how a Vegas slot machine has a programmed rate of coming up cherries, set by the casino. In the body, these odds are controlled by the laws of physics, which govern the amount of energy it takes for a mutation to occur. Exactly where that happens in our DNA is random, but the fact it's happening is not random.

We would all melt into the floor if those errors weren't constantly repaired. Molecular repair mechanisms are always at work, catching mistakes and fixing them, like a team of proofreaders fixing the work of a dyslexic writer. They're not perfect, but most of the mutations are silent, hitting a spot in the code that is ignored anyway. Genetics is more like writing than computer programming, in the sense that you can still read prose even if it has an occasional typo.

It's this mutation rate that allowed us to both evolve into humans and survive as humans. Mutations helped us survive the Black Plague in the Middle Ages by engineering us to absorb more iron. But now that same mutation damns us, as we overaccumulate iron in the liver over the decades of life. That mutation also makes some people immune to HIV. A similar story can be told of mutations that helped us survive malaria, survive frostbite, digest grains, run long distances, live at high altitudes, and far more.

But some of the time, these uncaught errors hit certain base pairs that govern critical cell functions. That may sound like it causes cancer, but it's actually normal. Unable to function, equipped with broken protein machinery, the cell just dies, the vast majority of the time. No big deal. It's when the protein machinery is altered in ways that *don't cause the cell to die* that cancer-causing mutations cause trouble.

Cancer is a lot like Facebook. It's a network. There are hundreds of pathways in a cell that can contribute to cancer. But some of those are more well connected than others. If you've got a cell pathway that doesn't have a lot of friends, it takes many mutations on those genes to trigger cancer. But if you've got a superconnected pathway, even a single mutation can flip its genetic master switch. On Facebook, a false claim can spread rapidly if it's posted by someone with millions of followers; it gets such a head start on truth-correcting folk that it takes on a life of its own. In the same way, well-connected master switches trigger cancer to take on a life of its own. Cancer cells are both fast-growing *and* fast-mutating. Mutations drastically pick up speed.

Cells have a truth-correcting mechanism, too. It's got a funny name: p53, or TP53. The 53 in its name simply describes its size. *T* is for "tumor," and *P* is for "protein." It'd be like naming me "IH143," for Indian Human 143 pounds. TP53 scours our cells, looking for signs of cancer, like a superhero. If it sees problems, it messages its friends to either prevent the cell from replicating, or kill the cell. TP53 keeps us all alive, every day. We *all* have cancerous cells forming in our bodies, regularly, every day. We don't progress from cancerous cells to "cancer" because TP53 catches them. Elephants rarely get cancer because they have twenty copies of TP53. Humans just have one.

But when a mutation happens to TP53, it's like Anakin Skywalker turning into Darth Vader. TP53 morphs into its other forms and turns traitor on the network of friends it's connected to.

My mom started feeling a pressure in her side about a year after the hysterectomy. Pressing her hand into her side, she felt it for the first time. A hard mass. "What is this?" she thought. The biopsy results came in while she and my dad (a retired professor of oncology at UCLA) were driving from LA to SF to visit me and my family. Her oncologist explained the mass was stage four malignant endometrial carcinoma. The cancer had escaped her uterus and settled into her abdomen and was growing again. The doctor said eighteen months left. Fuck.

A tumor mass is an evolution of cancer that involves many switches and misdirections to grow. The first thing it needs is oxygen. As the cells divide and reach about a million in number (about a millimeter in diameter), they cannot get the oxygen in and waste products out. This natural size limit to growth can keep tumors small for years, even. But eventually the cancer finds a switch that turns on angiogenesis, the development of blood vessels and capillaries to feed it. So it begins to grow. Next it needs space. The cancer cells begin to secrete proteases like collagenase, which breaks down the extracellular matrix surrounding it to soften the tissue and create room around it. Now it has supplied itself with energy, waste management, and space. Next, it needs to fight the immune system. Tumors have numerous defensive weapons. One of the most powerful is how tumors camouflage themselves against our immune system. The outer wall of a tumor is packed very tight, not allowing T cells and natural killer cells entry. They have a marker on their surface called PD-1 (programmed death) that tells T cells and other immune cells to disregard the cancer antigens our immune cells see. It has been recently discovered that some tumor cells secrete communication packets (called exosomes) into the bloodstream, loaded with PD-L1. These exosomes go to the lymph node, where they tell the whole immune system to stand down, making the tumor invisible to our body's defenses.

We needed to figure out what to do next for my mom. The surgeon said cut it out immediately. The oncologist said treat it with chemo first. I find it grimly amusing that experts will always favor their expertise.

After thirty-five years, chemotherapy is still our first-line defense against cancer. Cisplatin, the most common and best chemotherapy recommended to my mom, acts by interfering with DNA replication. The platinum molecules in cisplatin force massive damage to DNA after replication, inducing programmed cell death. The hope is that the chemotherapy will kill the fast-dividing cancer cells before they kill the rest of you. The common side effects of hair loss and appetite loss are from the death of the faster-dividing hair follicles and mucosal cells. The irony is that chemotherapy can also drive evolutionary mutations as the cancer tries to survive. If TP53 mutates, now the cell death pathway is switched off. The tumor becomes resistant to chemo. So, sometimes, chemo can make cancer *more* dangerous. Despite this risk, we decided to give cisplatin to Mom first. Surgery to physically cut out what remained came second.

The morning of the surgery at UCLA's Ronald Reagan Medical Center, we all woke at 3 a.m. It was warm out. LA. Mom was first on the schedule. We always ask for the first slot if anyone in our family is getting a procedure done. It's like a family rule. Fresh surgeons make better decisions. I was terrified. I didn't show it. We loaded up two cars to drive to UCLA. Dad, Mom, and sister in one car, me and Krissa in the other. With us was a week's worth of stuff. Clothes, sleeping bags, camping pads to sleep on. Our family doesn't leave anyone's side.

As we checked into the waiting room, everything was dark and quiet. Except my heart. It was pounding loud and slow. Boom. Held Mom's hand. Boom. I looked at my sister. She was a wreck, near tears. Mom was tough. Defiant. They took her upstairs to get prepped for the surgery. We set up a base camp in the waiting room, blankets out on the ground.

The surgery was scheduled for five to six hours. It was long. They weren't sure what they would find. They wanted to get everything they could see. They said the chance of a colostomy bag was fifty-fifty. The chance of death, of course, wasn't ruled out. They fucked up the epidural over and over. I wanted to scream until it finally slid in. I wanted to take her place.

The surgical team finally arrived. I asked the surgeon to send me my mother's tumor so I could study it at our lab. I wanted a piece of the enemy. To study. To find weakness. The doctor agreed, and like that, they wheeled my mom down the hall.

I went downstairs to base camp and began the long wait. An hour went by. Boom. Boom, Boom. My heart was beating faster. The TV screen showed all the patients' procedure statuses. "In Surgery." "In Recovery." Like staring at the arrival screen at an airport, waiting for your love. At first, the delays trigger anger. But the longer the delay, the more it's fear. The fear the TV will suddenly display "See Concierge" and someone will take you into a private room. You pray, god or no god.

Boom Boom Boom Boom.

At hour seven I went out into the blinding LA sun, past the palm tree, around the brick wall, and began to sob and scream.

Collecting myself, I walked back into the room and held Krissa's hand. She understood and said nothing. Krissa had just become pregnant with our first child. Mom needed to meet her.

Mom's status on the monitor blinked green.

The surgeon came down and we spoke. He told us the operation went well; they removed a large mass and found several smaller nodules that they also removed. They were able to spare her colon, so no colostomy bag. She was going to be OK for now. We couldn't let our guard down. We still had a war on our hands. I asked about the tumor sample. He said he had it put on ice and sent overnight to San Francisco.

Charly Chalawan, a friend and chief science officer of cancer therapeutic company Synthex, received the sample here and prepped it for storage. It was still black from the chemotherapy burns. Many cancer cells died as they were prepped for storage. That was a good sign. We took the prepped tumor sample, put it in a small vial, and placed it in the liquid nitrogen freezer for safekeeping. I wanted to peer inside the genetic soul of the cancer that was killing my mother. Finding the actual mutations driving this cancer would help us understand the best way to treat it. We took the other half of the tumor and sent it to Foundation Medicine for sequencing. What we found was scary.

A total of nine mutations had developed in her abnormally fast-growing tumor.

CTNNB1
EGFR
KRAS
PTEN
ARID1A
CUL3A

JAK1
PIK3CG
PIK3R1

Thankfully, I didn't see p53 on the list. But there was plenty to worry about. Scientists today have a good understanding of what each of these mutations does. The mutations told the story of her tumor, mechanism by mechanism. Each of these nine mutations made my mom's cancer faster growing, more at risk of metastasizing, and more resistant to chemo.

I was certain that there was still a substantial amount of cancer cells still strewn about in my mom's abdomen. We were hopeful that fresh rounds of chemo and radiation therapy would beat back the small pockets of cancer left behind, invisible to the naked eye and overlooked by the surgeon. We also started my mom on the newest blockbuster drug science had created. Immunotherapy. This is a term used to describe approaches to retrain your immune system to defeat cancer's defenses and kill the cancer. Keytruda is an antibody developed by Merck Pharmaceuticals that binds to PD-1, neutralizing its camouflage, so that T cells can attack it. We gave her hormone therapy as well. There is no one cure, but we were using a kitchen-sink approach to slow or stop the tumor growth. In the absence of complete eradication, we hoped for at least a standstill in growth.

And that was what we got. My mom is a tough woman. She endured new rounds of chemo, targeted radiation therapy, hormone treatment, and immunotherapy. The cancer cells are still there, revealed in bright red dots on her quarterly PET scan, but thankfully they are not progressing in size. Dots, not golf balls. It has been almost four years. The war is still raging.

Every day I worry about mutations. Mutations that make the cancer stronger than my mom. Mutations that overcome her body and our drugs. A mutation that will make the red dots grow and grow and spread until the PET scan is stained a solid red. It drives me to fund new approaches to treat this terrible disease of mutations and experimentally make progress. We have four companies that target cancer's well-connected master switches, previously considered "undruggable." Each has a unique approach. We have another company that attacks cancer's nutrient supply lines–depriving cancer, literally, of its food supply. We are always looking for more. We will not stop.

I am stalking cancer, but cancer is stalking my mom. Knowing the mechanisms down to the genetic level, however, doesn't offer me any mechanisms for managing my fear, or consoling my family with their fears. I don't know how many years my daughters will have with their grandmother. To cope with that uncertainty, I have no expertise. I don't have a plan. Just enjoy every day we can.

15

Memory Loss Can Be One of Effects of Lightning Strike WMTW Channel 8 (ABC)

It all started when Arvind couldn't remember what he'd had for dinner the night before.

I've always had a childlike fascination with memory. What is it? How does our brain store it?

Somehow, in all the neuroscience I've read and written about, this fundamental tidbit hadn't been answered. My mental model of the brain always felt a little compromised by not understanding this elementary building block. It niggled at me.

At IndieBio, we are always going after the brain. We've funded stem cell therapy, brain-computer implants, molecular therapeutics, and neuroimaging AI for neurosurgery. We host "Brain Dinner" salons with neuroscientists and neuro-inclined investors, and cohost events on using psychedelic drugs for depression and PTSD. Somehow, despite interacting with all those experts, I hadn't learned what memory *really* was. Or let me restate that more precisely: The explanations I'd been offered always felt incomplete. I didn't really need a deep understanding as much as I wanted a basic but *satisfying* understanding.

Last week, Sam Rodriques came by to brainstorm a company. He was a young, star theoretical physicist who, just four years ago, decided to learn how the brain works. He's invented several new technologies to elucidate brain function, in areas ranging from nanofabrication and single-cell RNA sequencing to quantum computation. Sam is deeply focused on learning how the human brain works at the level of each coordinated neuron, not by brain region or nerve bundle. It's here, he believes, that we will tease out the future understanding of consciousness.

As we began, I mentioned that Arvind couldn't remember what he had for dinner last night. He'd told me this earlier in the day.

Arvind added, "I texted my wife–so I know, from her, that we went out for Chinese–but I just don't have a memory of it."

"Probably a glia ate it," Sam blurted, adding, "It's crazy, you can actually watch under a microscope, glia eating memories."

Arvind jumped out of his seat. "What!? You can *see* it?"

Sam laughed. "You can actually see the memory *in* the glia's stomach." He revised his remark to make it more technically accurate–the microglia eat the synapses, and each synapse is a memory.

This was *weird.*

I'd never heard of being able to see memories under a microscope, or things eating memories. If you could eat them, that meant they were a physical thing, not just a current in a wire.

We all started furiously drawing on the whiteboard wall.

We spent an hour with Sam that day, and another hour a few weeks later. In between, I'd been reading everything and talking with a number of neuroscientists, and I'd realized something. The reason nobody ever gave me a satisfying answer to how the brain stores a memory is actually quite simple: *They didn't actually know.*

In other words, it wasn't just Arvind who didn't know where his memory of Chinese food was–*nobody* quite knew where it was.

But they are now tantalizingly close. And what they've learned, just in the last two years, paints a quite different portrait of the human brain's workings. It's actually so wild that most neuroscientists are not familiar with it yet, though maybe they've heard rumors. So here's the Top Ten List of crazy new stuff about memory that I learned while hunting for Chinese food in Arvind's brain.

Top Ten List of Crazy New Stuff About Memory That I Learned While Hunting for Chinese Food in Arvind's Brain

1. Neurons have a second "genetic code," a cognitive sequence code, which they write and read.
2. Brains don't "store" memories, they *grow* them, physically.
3. These growths are highly dynamic; they can change shape in a matter of seconds or minutes, or get wiped out in a matter of hours.
4. You can learn only so much at one time because our neurons run out of building materials to grow more memories.
5. Our rich ability to imagine things in the mind's eye is the result of a virus.
6. In order to record all these memories, we gave up something that fish have, which is the ability to pass cognitive sequences down to our children.
7. Imagination is the reversal of memory.
8. In the future, we will be able to read these cognitive sequences.
9. We will never be able to upload our brains to computers by the laws of physics as we know them today.

Let's unpack those.

1.

Classically, neurons are both circuit and cell. Where neurons touch, neurotransmitter chemicals flow, allowing one circuit to trigger another. So it was often said that the brain was both electrical and chemical.

Since a neuron was a live cell, of course it needed to have a nucleus, and in that nucleus would be genetic machinery, same as any other cell. But this was hardly the most distinctive feature of neurons, so the genetic machinery of neurons was usually the last thing you'd hear about, if you heard about it at all. The normal stuff that nuclei do–divide, replicate–wasn't needed in neurons throughout life. They don't replicate. They can last a lifetime.

But a new paradigm is needed. Electrical, chemical, *and* genetic.

Imagine for a moment what life would be like without conventional memory. You would be constantly in the present, reacting to your internal urges and what's around you, like what Buddhists describe as having the mind of a candle in the open window, dancing in a new direction with every shift in the breeze. This is how many organisms live. Their electricity runs just fine. Their chemicals run just fine.

The difference between you and a person living in that present-only state is largely due to the operations of genetic machinery.

Think of it like sheet music being created. As a neuron fires, *zap zap zap*, so too do certain genes in the neuron, *zip zip zip*. The pattern of gene activity is recorded into RNA. So now this chunk of RNA has (indirectly) the neural firing pattern encoded. These cognitive sequences can be copied, moved around the cell, and used for various purposes, including "replaying" the memory by retriggering the neural firing, using the same mechanisms in the reverse direction, the same way sheet music can control a self-playing piano.

But that would be a *really* simple memory, because it describes only one neuron.

Earlier this year, a UCLA scientist transferred this kind of memory from one snail to another, by transferring the RNA cognitive sequence. The first snail had been shocked; the second snail had not. But with the transferred memory, the second snail acted like it had been shocked before.

But a snail has only two neurons. One says, "I'm hungry!" and the other says, "I smell food over there!" Humans have 100 billion. And to coordinate all of them, our neurons have some special capabilities.

2., 3., & 4.

Memories survive for different lengths of time, from seconds to a full lifetime.

As Arvind was eating his Chinese dinner with his wife, neurons in his brain that took in the smell and heat and taste of the food would have genes firing. A little chemical tag gets attached to these genes, then more genes get involved, and within a few seconds you've got some cognitive code.

At that point, the memory is not going to last very long, even if it's encoded. Because, again, this description is just of a single neuron–it's not even a whole thought. To get stored for the long term, what it really needs to do is connect up with all the other neurons that were also firing at the same time during dinner. For instance, Arvind's memory of his Chinese food should also be connected to a category neuron for his wife, Krissa (one that all Krissa memories connect to), and it should be connected to a category neuron for *ma po* tofu, because he's had a lot of *ma po* tofu in his life, and he rates all Chinese restaurants in America on their *ma po* tofu.

That connecting step can take seconds to minutes, and I found this to be pretty wild. At one end of these neurons you've got dendrites, looking like tree branches. Off the branch of each dendrite will grow a fiber, or a spike, or a spine. Scientists use different words to describe them. At a certain distance they look just like bumps or buds, but zoomed in they kinda resemble mushrooms. Each dendrite will have around a thousand of these buds on them, so your brain could have up to a hundred trillion of them, total. As Arvind sits eating dinner, his Krissa neuron and his *ma po* tofu neuron will also be activated, and the sensory neurons will grow a spike out to grab on to the axons of those also-lit category neurons and create a synapse.

Some people would call this "neuroplasticity," and they're wrong. "Neuroplasticity" refers to retriggering the adaptive state of a baby's brain, during which neurons differentiate and grow and send axons crisscrossing everywhere. That's *building* a brain. This is something else. This is the brain *in operation.*

That our brain has to rapidly and physically construct these spikes in minutes–and tear them down when they're unused–means that our brains are highly active construction zones at all times. It's a *lot* more complicated than a computer with fixed wiring. The brain is simultaneously a memory-making factory. The spikes are complex. These buds are only about a micron in size, like the size of a small bacteria, but they have vesicles to ferry RNA and neurotransmitters, and machinery to make more. They're flexible and dynamic. These buds will create proteins that then sit outside the bud, almost camped on the synapse, regulating it. But if you come back in a few hours to that part of the brain, 10 to 20 percent of the buds can be wiped away.

That was what Sam was describing, when he said a glia cell ate Arvind's dinner memory. Glia are the cleanup crew.

Sometimes, when people are struck by lightning, they gain the ability to rapidly learn new things–such as playing the piano, or a new language. And if you want to improve your memory recall, you can now use a brain zapper that sorta warms up your brain, changing the resting state of neurons, so that it takes less energy to fire them. The inherent problem with cramming for an exam is that at a certain point, your brain cells run out of construction material.

7.

We use memory to recall the past. We use imagination to envision the future. Across the world today, half the people feel the world is changing too fast; half the people feel it's not changing fast enough. One could say that the former want the world to be more like they remember it; the latter want the world to be more like they imagine it. In this sense, memory and imagination are locked in eternal battle.

Imagination is quite amazing–a leap forward from mere movie-dense memory. You can imagine things that have never happened, but you can also imagine them with movie-like snippets, using memories or recombinations of them. We wrongly characterize imagination as resembling dreaming; technically, this isn't accurate at all (different lobes of the brain are involved). Imagination starts in the prefrontal cortex and seems to use the wiring of memory, but backward, going to the parietal and then down into the visual.

Let's practice. I'm going to ask you to imagine something. Conjure this image in your mind's eye.

I want you to imagine a really big high school gymnasium, perhaps a basketball court lined on its hardwood. Now imagine this gymnasium stretches off into the distance in both directions–it's almost infinitely big. Now fill this gymnasium with pianos. A *billion* pianos. A few pianos in the foreground are playing music. As these pianos play music, pages of sheet music are being created automatically on the music rack, recording and storing the notes, almost like an office printer churning them out. And as the music continues to play, these pianos change shape slightly, becoming longer, adding keys, having a thicker soundboard.

What you've just imagined–that image–would have memory. But it would not be able to imagine new music.

5., 6.

I just don't think most people are aware how much of evolution was driven by viruses, not mutation. Viruses integrate their DNA into the host's DNA. Some of the time, this kills the organism. But sometimes those edits made the organism better–and gave it wholly new powers. It's been estimated that at least 25 percent of mammalian DNA came from viruses, not spontaneous mutations—with estimates as high as 80 percent. Even if you just look at what separates humans from chimpanzees, 30 percent of the 120 million base pairs that are different originated with viruses.

Viruses can attack every part of the body, including neurons. Even today, viruses may be at work behind the scenes in a number of neurological disorders. Viruses can become dormant and reactivate decades later. Take the seemingly innocuous condition of infant roseola, which causes a rash to temporarily break out on babies' faces. Almost every baby gets it, then it passes. It was considered harmless. But it can get reactivated, and its reemergence later in life has been associated with multiple sclerosis, chronic fatigue syndrome, Alzheimer's, and depression.

Fish can pass memories between generations. But maybe the better way to say it is that fish can pass experience learning through generations. If a fish learns something really important during its life (I love Oregon!!), its spawn of children inherit that knowledge. Humans get wiped clean during embryogenesis, and a few more times during childhood, with only a handful of exceptions. But for this trade-off, we have a much richer memory during life.

It appears that we owe our complex memory capability to an ancient virus that integrated some very special DNA into the minds of our evolutionary predecessors. Something happened there that was *not* normal. Inside the nucleus of a neuron, this gene, ARC1, builds a very unconventional protein. It's a protein that looks and performs like a virus. It carries genetic code and builds for itself a virus-like shell, which allows it to travel across the synapse between dendritic spikes, where its genetic code can be taken up by the next neuron. It's impossible to make durable memories without it.

Science isn't ready to declare this neuron-to-neuron genetic transfer is a memory, per se. Rather, just that it's the key to connecting the network of neurons that fire together when a memory is first created, and later brought back to life in the mind.

As yet, nobody really knows why genetic code is passing between neurons that are bound together by memory. But it's happening, in your mind, this very moment. I still don't quite have the answer that satisfies my curiosity, but at least we found Arvind's dinner.

8., 9.

All this brand-new science presents a gigantic obstacle for the people who dream of one day uploading our minds to a computer. We work with these people...they're friends of ours...what they're doing is daring. But their grand plans didn't account for the discovery that our mind, our consciousness, would be (at least partly) stored as genetic code. And that as we recall memories and intertwine them with new thoughts, we're entangling all that code.

What they were counting on was the conventional idea that neurons are more like electrical wires–more like a silicon circuit. So the field has advanced remarkably on that front, inventing electrophysiological arrays that slide electrical probes (or light-sensitive actuators) down between axons. And with these devices, we are just decades away from being able to *alter* consciousness by stimulating one brain region or another.

But altering consciousness is a far cry from creating consciousness. A lot of things alter consciousness. Alcohol. Emotions. Suffering. Learning.

Consciousness–that voice running through our heads–won't be decoded until we have the capacity to capture and read the genetic code that is written with every thought.

Which gives me an idea.

16

U.S. State Set to Outlaw Calling a Veggie Burger a Veggie Burger The Guardian

Arturo and David wanted to save the chickens.

George and Matthew wanted to protect the rhinos.

Uma wanted to spare the cows and pigs.

Dominique and Michelle wanted to rescue the sharks.

They all wanted to save Earth.

In another time or another place, nobody would have cared what a scientist wanted. But something different happened when the barriers to creating a company were drastically lowered for scientists.

They met at conferences and communities like New Harvest. PhDs in biology teamed up with animal rights activists to conceive of a new way to make food without harming animals. They applied for government grants, but the government had no idea what to do with applications for biotech in food. These applications are reviewed by other scientists who flat-out did not understand why this would be an area to apply biotechnology.

So they reached out to venture capitalists. They were met with blank stares. Biotech was for making drugs. VCs couldn't wrap their heads around biotech in food.

When I started IndieBio in 2015, I had no idea what would walk into the front door. I was breaking every rule of biotech. Scientists aren't entrepreneurs. Biotech needs tens of millions of dollars. Biotech means making drugs. But something else happened. A different breed of people walked through the bright orange door of our first lab in the Dogpatch of San Francisco. Renegades. These scientific revolutionaries came to me because they had nowhere else to go. IndieBio was just starting and had nothing to offer except a barebones lab and a pittance of money.

Sitting on my desk was a one-page plan. Yeast is the common name for *Saccharomyces cerevisiae*, the fungus that bakers use to make bread rise. It belches CO_2 as it eats sugar, and when it doesn't have much oxygen, urinates alcohol. Clara Foods, a company that did not yet exist, proposed to inject into yeast the chicken genes that make egg proteins. Once floating in the cytoplasm of the yeast, the protein machinery would read the egg white genes and make the egg white proteins. After the yeast broth was thick with egg whites protein, they proposed to strain and purify it until it was a pure white powder. This powder would then be remixed in proper ratios with water into a perfect-tasting, perfect-cooking egg white. Not an egg substitute. Real egg whites–without the chicken.

I knew how farm animals were treated. I'd seen slaughter. But until that moment, I'd also willfully ignored it.

Mahatma Mohandas Gandhi once said, "The greatness of a nation can be judged by the way its animals are treated." Gandhi's concern for farm animal welfare was not secondary to, nor a corollary of, his philosophy on nonviolence–it was fundamental to it. He felt that nonviolence would lose its moral coherence if it applied only to political protest. He spent thirty years advocating for more humane treatment of farm animals, and he always forbade their slaughter.

It was a very simple question for scientists to ask, one hundred years later. In order to eat meat, is there another way?

I was in for the ride.

Rhino horn is made of keratin, the same stuff in your nails and hair. Keratin, similar to DNA, is a superhelical filament and grows from your nail bed at the astounding rate of a nanometer per second. Which I find to be contradictorily fast and slow. A rhino horn is as unexceptional as your toenail. Except in Vietnam and a handful of Asian countries. In those countries many believe that in addition to keratin, rhino horn contains magic. A magic ingredient that will cure cancer, detect poison, and make your penis hard as a rhino horn. Since magic is not detectable by today's scientific instruments, George, a chemist who built a successful shampoo company, and Matthew, a serial entrepreneur with a love for rhinos, had an idea. Sitting across from me in our first lab, they explained: "If we could make a synthetic rhino horn in the lab that was identical to the rhino horn found on rhinos, we could flood the market with the cheap stuff, crater the market price, and save the rhinos."

I couldn't ignore the rhinos, either. I had no idea if it would work or if people would buy it. But I built IndieBio to try and see, not think and predict.

Rhino horns were going well. They built a custom 3D printer and printed rhino horns with keratin powder as the material. To complete the ruse, they blended in rhino DNA fragments they amplified on the PCR machine in the lab.

Meanwhile, egg whites were not going well. Their first attempts were a terrible clear slime with no taste that you didn't dare swallow. We didn't know why. The proteins were pure, they had purchased them. The ratios were right. On a digital scale with three zeros, they measured them. Something was missing.

The rhino guys were experts in minerals. They sat next to the chicken guys in the lab. "You're missing minerals, dude." Calcium, or CA++, tastes slightly metallic and is present in so many foods. It also acts as a stabilizer between the proteins, which probably improves the texture. Out of other ideas, the chicken guys gave it a try. The result tasted like eggs.

So far they had made a few micrograms. Only nine trillion times more to go.

Rhino horns were rocking. In our lab they used a custom-built 3D ceramic printer and keratin powder to produce a thumb-sized cone of rhino horn. It was reddish and hard. It looked real. If you sent it to a lab they would tell you it was real. Of course, no one agreed on what would happen. Investors didn't know what to make of it, so they decided to wait. Matthew is not the guy who waits for others. He created PembiCoin and ICO'd the world's first rhino horn cryptocurrency.

They raised less than a hundred dollars.

Alex and Nick wanted to save the cows.

Mike and Brian wanted to free the tuna.

"This is the DNA of woolly mammoth collagen." Alex held an Eppendorf tube between his thumb and index finger.

"Whaaaat?!" I said. "How'd you get it?"

"Public data sets. We just had the gene made and shipped to us. We're gonna make a gummy bear made of woolly mammoth gelatin as a proof of concept. By putting the woolly mammoth gelatin gene into our bacteria factory, it will spit out mammoth gelatin. Purify the gelatin and pour it into gummy molds."

"That's insane. I wanna try some!"

They ended up being orange. Looked exactly like gummy bears. Tasted like them, too. But I had just eaten the resynthesized proteins of an animal that has been extinct for ten thousand years.

Alex and Nick did their PhDs at Princeton, with Alex earning an MD at the same time. They were studying *E. coli* reproduction when they realized they could arrest cell division at a critical point and convert the cell's energy to make a protein. This would dramatically improve the productivity of synthetic biology. The only question was what protein to make. I was seduced by the potential of these zombified prokaryotes and thought about how many cow noses it takes to make eye cream and other products. I thought about all the cow noses and hooves that could be spared by making collagen in the lab.

The power of biology as a technology began to dawn on me like the sunrise on a planet new to sunlight. It powered a revolution in my thinking. We could remake the world.

Then there was Uma.

When I called Uma Valeti the first time, he and his wife were driving south out of Minneapolis on Highway 52 down to Rochester, where the Mayo Clinic is. Uma pulled off the highway near Cannon Falls, and we talked.

As an interventional cardiologist, Uma had spent years injecting cardiomyocytes into patients' hearts to increase their heart muscle. For almost a decade, he had been dreaming of bringing this technology to meat production. His idea was to grow pork muscle stem cells in an incubator to make a sausage. That people like you and I would eat. Instead of pigs. Ending slaughter with the incubator. Briefly, the company was called Meat Seed, but soon after getting to IndieBio, they changed its name to Memphis Meats.

As news of these food companies using biotech got out, it created confusion. Then when people got it, they laughed. It was funny! It was cute.

AbbVie and Merck and the other Big Pharma companies would visit me in the lab and chuckle when I would talk about the food applications of biology. "We have been making antibodies and other products with these methods for years. It doesn't scale easy! That's why these therapies are so expensive!" They chuckled politely because they saw the attempt to grow animal cells at scale as naive. And they are right. It is naive, but it still may work. I never understood why people laugh at the unknown. It is a beautiful place, not knowing. Learning through action.

Journalists started covering the movement. Conan O'Brien and the late-night crews got in on the act. How could they not? Conan said, "I would try a test tube hamburger. If it was 4 a.m. and there was no food in the house and I could make a hamburger with a test tube, I would do it."

Another lady in a BuzzFeed video said, "I'd eat lab-grown meat. I have no standards when it comes to food." These videos have millions of views.

Stephen Colbert did mock traditional hot dogs as well. He noted that 2 percent of (traditional) hot dogs contain human DNA. And the source of the DNA is unknown. He muses, "Could it be hair? Or a romantic moment at the sausage casing machine?"

Most VCs also gave me a polite chuckle when I showed them our food companies. Citing the same issues along with the general difficulty of growing at scale, and the fact it was just plain weird. It was all too risky, they told me.

Brian Spears heard the laughter of the scaling problem almost daily. So he went to NASA and found a way to put the cells into orbit. On land. He licensed a special bioreactor that enabled the cells to relax in a comfortable environment so they could grow faster and more densely. They held a public tasting two months later. There were reporters at the event laughing that it didn't taste exactly like pork sausage, but close. Others were stunned at the rapid progress everyone was making.

By funding food, we reinforced the biotech community's view that the IndieBio accelerator was, in fact, crazy. But who cared? We were laughing as well, not at them, but at what was possible and because we were having so much fun.

Three young women wanted to use algae to make shark fins for shark fin soup. Dominique was a shark biologist. Michelle was a food scientist. Jennifer was an entrepreneur. They wanted to save the sharks.

Everyone is horrified by shark fin soup. To make it, you must catch a mature shark in open water. Hauling the top predator onto the boat, you take a long knife with a hook on the end and slice the dorsal fin off the shark from front to back. Then roll the shark back overboard because it's way too big to lift. It's like a beheading. Without its dorsal fin, the shark is sentenced to an agonizing, slow death by exhaustion and drowning.

No one should eat shark fin soup–but millions in China were serving it at special occasions. As with the rhino horn, it raised the question of making a substitute product. Would it sell? Would it increase demand?

Yao Ming, the former NBA star–and arguably the most recognizable athlete in China–began a campaign against shark fin soup. Many others followed. It became uncool to eat any form of shark fin soup. The status of shark fin was replaced by the shame of shark fin.

New Wave Foods found this out as they tried selling their algal shark fin. The people that woke up to the cruelty of the dish wanted nothing to do with it. New Wave Foods quickly pivoted to shrimp. They want to save the shrimp.

Applications came pouring in. Mice. Udders. Foie gras. Lion. Kanye West. Yes, *Kanye West* burgers, from a single cell of Kanye. I'd never fund that–but the James Franco salami...was interesting. Why salami? Was he better cured? Did he need a cure? I never interviewed the founder to find out.

I was in a meeting room in New York talking about helping make the city become a world biotech hub. State representatives and private equity funds were all on board and loved our approach. Concluding the meeting, we were all laughing, and we agreed to help them. They said great, "but leave that food stuff in California."

Kimberlie wanted us to eat mold.

Matt and Inja wanted to save dairy from itself.

Our little revolution was no longer ours. People began to notice. Powerful people.

The cattle industry, after ignoring the initial companies as oddities, began to organize to discredit them. *Beef* magazine (that is actually what it is named) called the fake meat companies "fake news that needs to be stopped." They also wrote an article titled "We Have Met the Enemy and It's Fake Meat Companies."

Arkansas state legislator David Hillman had had enough. On behalf of the cattle industry in Arkansas, he drafted and passed a bill to ban the term "veggie burger" because it confuses consumers. Maybe in Arkansas. He inspired twenty-five other states to follow in his footsteps.

The second salvo came with a petition to stop the label of "clean meat" for meat grown in a lab. Labels are important. If lab meat was clean, what was meat from slaughtering cows? Dirty.

The only thing to do was fight back. With reason. With nonviolence. So, Uma, CEO of Memphis Meats, and Brian Spears, CEO of New Age Meats, went to Washington for a hearing on the subject. Together with the meat industry, they jointly agreed to relabel "clean meat" as "cell-based meat." The meeting was a landmark of cooperation between the incumbents and innovators. Maybe it is a truce.

Next came the fight of safety. Meat industry officials exclaimed, "How do we know it's safe?" I took a lesson from Dana White. Yes. Dana White, CEO of the UFC, Ultimate Fighting Championship. I had attended a talk he gave at the Stanford Graduate School of Business, where he was being interviewed on how he grew the UFC from $2 million to $4 billion in twenty years. He said something that stuck with me. "We ran towards regulatory. We were in the dark ages. The cable networks and pay-per-view wouldn't carry us because we were illegal. We could never get big without the commissions and legislature getting on board." Same for cell-based meat, I reasoned. Safety is the number one question on the minds of meat consumers. How would they know meat from a lab is safe–or even safer than meat from cows? By embracing the FDA and USDA to prove it. Though it may take longer to get to market, when it does, it would eliminate a source of real concern. So we counseled our clean-meat companies to get regulated. Run to Washington. Share with the public. Not just clean. Open.

After another meeting in Washington, the founders of the cell-based meat companies agreed to be regulated by the USDA and the FDA. "We wanted to be sure it was safe anyway, so why not get official credit?" Uma said.

This was no longer a fight between a trillion-dollar meat lobby and a startup here and there. It was snowballing into a full-blown reshaping of the industry. No wonder they were going after us with all their might.

Even the rhinos were getting a hard time. Pembient, our rhino horn company, was getting massive publicity. Front page of Reddit. CNN. NBC.

ABC. TechCrunch. You name it. Then the controversy exploded. The Humane Society declared Pembient's solution inhumane. They argued it would increase demand for the real thing because more people would use it. The economists had a field day with this, spawning several papers from respected places like the International Trade Centre. There are now only two northern white rhinos left and they are both female. The species is functionally extinct.

David wanted to save the little fish.

Matias, Pablo, and Karim wanted to end the use of animals for food.

Sam Harris, an influential blogger, tweeted a poll on February 1, 2016: "If cultured meat is molecularly identical to beef, pork, etc. and tastes the same, will you switch to eating it?" The reply stunned me. Eighty-three percent of almost fifteen thousand people said they would. This was a turning point. Once there was some market validation, however thin, investors and scientists began to take this movement seriously.

Meanwhile, I hit Sand Hill Road (the road with all the fancy VCs lined up in a row) with a presentation on the future of food so they understood what we were doing. It started with a painting of a caveman in a loincloth spearing a bear while the bear was mauling his friend. It was an important slide because it showed the importance of protein. It was so important to our diet that we used to be killed trying to get it. Now it is so easy to get, we are killing our planet for it.

This message began to win the VCs. Without funding, the companies were dead. IndieBio is an accelerator offering $250,000. While we follow on and invest millions in our companies, they would need hundreds of millions of dollars to deliver their promise. To generate interest we gave (and still give) tours of our lab where VCs can see where and how this future protein is made.

In the end, chickens did better than rhinos. Scott Banister, PayPal mafia member and Uber seed investor, wrote a check for Clara Foods.

Scott is a vegan. So were the Partovi twins who coinvested with Scott. Soon after that, Google Ventures and Ingredion invested tens of millions of dollars into Clara to help it scale. As of writing, the company is worth more than $125 million, four years after they made their first meringues in our tiny lab.

I walked into a kitchen in North Beach, San Francisco, where I was greeted by a very happy young man. "Hey, Arvind! Please take your badge. No photos please. We will be starting in a moment." This was the historic first public cooking and tasting of Memphis Meats' meatball and duck. Next to me was a *Wall Street Journal* journalist who was eagerly taking photos. "Quiet on the set!" The cameras were rolling, the meatball began sizzling. The smell was incredible. Like a meatball I've had a thousand times growing up. The chef carefully lifted it out of the pan and gently placed it on top of a volcano of fettuccine. The taster, also from the *Wall Street Journal*, picked up her fork and knife and cut a quarter of the meatball off. She speared it and held it up. With the slightest of hesitation she popped it in her mouth. The entire room held its breath. We could hear her chew. You could see a wave of relief spread over her face from forehead to chin as her lips spread into a smile. "It's good," she said with genuine surprise. And an entire future industry breathed out in relief. Steve Jurvetson, who invested in SpaceX, led their series A round of $16 million. He was joined by Bill Gates, and Tyson, *the largest meat processor in the country*. The meat industry had declared, with hard cash: If you cannot beat them, join them. They are now worth north of half a billion dollars.

In the rest of the country, veggie burgers became the focal point of megatrends colliding. Millennials could eat less meat to fight climate change. Boomers could eat less meat to fight heart disease. Veggie burgers could be made with fermentation technology that "bled" heme, the molecule that makes blood red and tastes iron-y. Beyond Meat filed to IPO. In a fancy hotel in New York City, an investor asked me if they should invest in the IPO. Sipping a cappuccino, I told the group I had no idea. I was not an investor in Beyond Meat and I am not a financial advisor. But I did tell them this. There was no way to invest in the massive millennial movement to eat less meat and vote with their dollars to fight climate change other than the upcoming Beyond Meat IPO. Beyond Meat would be the only company that was singular in that thesis. Their stock quadrupled on the first day. I got text messages and emails from VCs–the very ones who had said I was crazy–now congratulating me on our food portfolio. The gold rush was on.

Stem cell engineers are suddenly seeing hot demand for their skills. Venture capitalists are fearful of missing out on the next big food company. Software investors are asking about stem cells and fetal bovine serum. Journalists are writing thousands of news articles. I now receive industry reports on synthetic meat from major banks and data services. As of this writing, there are fifty-nine cell-based-meat companies, and double that many alternative protein companies. They are all getting funding.

Nonprofits like the Good Food Institute are organizing conferences around cell-based meats and alternative proteins, with thousands attending. I regularly receive industry reports on the future food and agriculture industry. It makes me happy.

If you are still leading a revolution, there is no revolution. Even if all this doesn't work, even if people eat animal-based meat twice a week, I am happy that people are thinking about the impact of their decisions. On themselves, the planet, and the animals. I know that is now true for me.

Cargill, ADM, and other large-scale fermentation companies are backlogged and inundated with requests to scale products. Foreign governments like Singapore are looking at fermentation as a new way of securing the national food supply.

The world is fundamentally different than it was five years ago. Billions of dollars of market value has been created by alt-protein startups and billions more lost by Kraft Heinz and other incumbent brands. Revolutions aren't just zero-sum. They reframe and reorganize the world along new lines. Americans no longer care for the thirty-second ad with a jingle but the values that the product upholds.

The real winners will be the animals. Spared the emotional stress of living unnaturally in a metal cage. Spared from their babies being taken from their mothers at birth to be grown into milk-producing calves. Spared from being born and bred as a meat-making factory, only to be a carcass. Another winner is the planet. Less water and land used, fewer forests clear-cut for feed, more wildness.

We've invested in more than thirty food and agriculture startups reinventing how we eat. We aren't slowing down.

Three years after Memphis Meats graduated from IndieBio, I drive Po over to their facility in Berkeley. We listen in to their all-hands meeting (the company now has more than forty employees) and we spend forty-five minutes catching up with Uma Valeti, reminiscing about the old days.

Then we head to one of the cold rooms in their small production facility. A freezer opens. Inside are rows and rows of frozen, vacuum-packed chicken breasts. They are thick, large, and almost ordinary. So familiar yet unimaginably pure. We glide into the company's kitchen. A defrosted breast has been sliced into hunks, then marinated in coconut milk, with a little soy sauce, some yellow curry. The aroma of turmeric, garlic, and ginger hits as we slide the chicken into the nonstick pan.

As the chicken sears and carmelizes, we talk about TV shows, then science TV shows. Peanut butter is mixed with the same spices, plus some lime, then a quick pickle is given to shavings of cucumber.

As we enjoy the satay, the talk turns to our kids, our families, our travels.

At this moment, it's not about revolution. The days of being laughed at are far from our minds. It's not about acceptance. It's not even about expectation or pressure to succeed. Instead, it's about something permanent, something we all love and know well–both ordinary and timeless–the joy of eating with friends on a winter's day, inside, warm.

9:36 PM
83%
Messages
Arvind
Details
We need to write a chapter about free will
about the role of free will in a future determined by the genome
everyone gets that wrong
everyone
but how do we tell that story?
on the plane back from New Orleans yesterday, the newest Jason Bourne movie was on
what does that have to do with it?
...
oh—I got it
Text Message

17

This Guy's $35,000 Cat Clone Looks Nothing Like the Original Futurism

In *The Bourne Identity* and its three sequels starring Matt Damon, the character of Jason Bourne embodies the timeless paradox between free will and determinism.

Bourne wakes up with amnesia, not even knowing his own name. He's a man of supreme decisive action, evading government agents and assassins, and yet even he has no idea what he's about to do. His actions look like free will, but his *mind* reveals otherwise. All his nifty escapes are programmed into his deep brain, his cerebellum, a result of special Treadstone training that he doesn't consciously remember, but which he can tap on demand, like a reflex. His free will is an illusion. He's a pawn. He goes on the primal quest to find himself, and that means learning who programmed him, and why. Decoding his past is the key to counterprogramming and freeing himself. He needs to know where the programming ends and the real man begins. Throughout the series of films, CIA intelligence operatives try to persuade him to quit his solo quest and rejoin the agency, ironically tempting him with the line, "You have a choice." He's never had a choice.

Though the Bourne movies are thrillers, the plot is eternal. Strip out the action and international intrigue, and Bourne is like any college student, using the separation from family to figure out who he or she *really is*, second-guessing the programming in which they were raised. Who am I, really?

Identity can be like unpacking a Russian doll. In search of who you really are, you remove the outer layer to reveal what's on the inside, then another, only to find that you are nothing but layers.

Free will has been debated throughout human history, and as long as we keep genetics out of it, I think we can summarize it in one paragraph.

The ancient Greeks endlessly deliberated whether the past controlled the future, and even though they agreed it did, they found some loopholes. Socrates believed man was controlled by his impulses, by his reptilian nature, but with self-control and reflection he could transcend his primitive nature. Plato said we coexisted outside our physical bodies in a transcendental realm. Aristotle felt the past did control the future, but reason could free a person from this trap. Monotheism came along and had its own free-will dilemma; if God was all-powerful and controlled everything, then how could you shame a man for his immoral act? The original sin, defying God, symbolized man's free will–but he would be punished forever for it. René Descartes solved this like a kind of Catholic Plato, with a dualist theory, that only on Earth did man have free will. John Locke was Socrates all over again. Thomas Aquinas echoed Aristotle; man's goals were programmed, but how he achieved those goals was up to him. Friedrich Nietzsche rejected all those guys and took the contrarian view that we were free if we wanted to be, but most people don't actually want to be. In the American century, the twentieth century, the concept of self-determination became a publicly espoused ideal. So the thinkers had a field day poking holes in it. Along came Sigmund Freud to say we were controlled by deep psychological wounds from childhood. The behaviorists taught us we were conformists. The sociologists had reams of data on how impossible it was to transcend socioeconomics and racism. Even on the best-seller lists today, Malcolm Gladwell reminds us of the hidden strings that control our fate, and Dan Ariely reminds us that we are not as woke as we think when we make choices. Even the physicists have now weighed in, saying that every nanosecond controls the next nanosecond, and it always has, so all of history and all of the future is just one long thermodynamic chain reaction, with free will a mere illusion.

Almost every single philosopher in history would agree with this fundamental statement: You are not who you think you are.

It was into this historic debate about identity, responsibility, and social mobility that genetics arrived some 150 years ago, like a giant meteor. And I'm going to make an unusual argument. Which is that the last 150 years of history would have been really different–our society and ideals and debates, all different–if we had actually understood *from the start* how genetics actually works.

Social philosophy always took genetics as determinist, as hardwired programming. And we thought that because we learned certain pieces first, those certain pieces were the low-hanging fruit on the genetics tree. But with far more pieces in place, today, genetics is primarily an argument *against* determinism.

If we had understood this 150 years ago, it would have had the opposite influence on all subsequent social and political theory. Rather than "Survival of the Fittest"–which wasn't from Charles Darwin; it was a philosopher interpreting Darwin in 1864–we might have had the opposite, "Survival of the Most Cooperative." Psychology might have run a very different course, because it wouldn't have borrowed the concept of "inherent" personality traits. Freud might have had nothing to debunk. We might never have engaged in the dialectic of *Nature vs. Nurture*, because we would have known that nurture permanently changes gene activity (that nature and nurture are one and the same). Rather than seeing life as a futile rat race for individual prominence, we might see life as a futile endeavor for collective well-being.

In other words, the sociopolitical history of the last 150 years is all due to a big misunderstanding.

Genetics was a determinist trump card. Determinists had "the code of life" on their side. Sadly, they still play this card, again and again, using it to justify hatred, racism, nationalism, and social inequality. Every time someone goes on Ancestry.com to find answers to the eternal question of "Who Am I?" they play with the same fire.

As I write this, I'm sitting at IndieBio in the Ivory Basement, at my desk, looking at the thirteen startups all working within sixty feet of me. Of those, twelve are actively altering life's genetic machinery. But they are doing it in twelve different ways, for twelve different purposes, hitting twelve different targets. When you see this on a daily basis, you come to understand that genetic machinery is insanely diverse in every dimension. One startup is using CRISPR, yes. One is using recombination. One is using viruses. One is using zinc fingers. One is using directed evolution. These are the classic ways to "edit" the genome's code. But the other seven startups are changing what genes get turned on or off, and the tools for this are limitless. One is using milk. (Yes, an ingredient of milk will turn key genes on and off.) One is using a hydrogel, which has physical properties that turn certain stem cell genes on. One is using a supercomputer to simulate a cancer pathway to turn genes off with drugs it's designing. One discovered a novel Cas12a enzyme in the extreme conditions of the high deserts of Argentina. One is turning human genes on and off with microbe poop. One has you eat a pill with DNA code in it.

On the macro level, the fact humankind can now control DNA, of our own free will, absolutely flips the tables on determinism. The era of genetic fate is over, and the more we master control of DNA, the more that genetics will be a matter of public policy decisions, not fate. That's obvious.

What's not obvious to most–but is *really obvious* down here in the Ivory Basement–is that "Genetics" and "Your Human Genome" are not the same thing. Your genome is just a subset of the overall genetic machinery going on in the body. Arguably, your genome is not even the biggest player in how genetics impacts life.

I'll offer a couple metaphors to reorient the mind, which I'll echo as we move ahead.

One metaphor is that the human genome is like the U.S. Constitution, but genetics is the reality of government and politics in society today. If an alien read the 1787 Constitution and its twenty-seven Amendments on its way to Earth, it could never accurately predict what politics and government is like today, across the nation, varying from state to state. The alien would land in Washington, DC, read the newspaper, watch a few political talk shows, attend some rallies, and send a message home: "It's a *lot* more complicated."

Another metaphor is this: Let's say that alien wanted to know what college students are reading today. To find out, it could go to the biggest college nearby, and look at all the millions of books in its libraries. And if the alien assumed that's what kids are reading, it'd be making a big mistake, because (1) that wouldn't tell the alien which books are actually being checked out or read; and (2) students are given a ton of assigned reading by professors, and only some of those books are kept in the library. Our genome is much like the college library. Just because it stores a gene doesn't mean that gene is on, just like a government program can't be effective without budget funding.

Almost everything in the collective conscious about genetics is wrong. Identical twins don't have the same DNA. If you clone a cat, it doesn't even look the same. If you have the genes for dark skin, it doesn't mean you have dark skin. Every cell in your body does *not* have the same DNA. Gene therapy does not make permanent changes to your genome. Sperm does not carry a "copy" of a father's genome. 23andMe cannot tell you what nations your ancestors came from. If you knocked out the gene for X, you very likely wouldn't get rid of X at all. The genes for intelligence do not determine your IQ.

I'm going to put this gently. The human body is chock-full of DNA, and only some of it matches your "constitution." We'd like to think all of it follows the constitution–that's the way we were all taught. And a good chunk of it does. But less than you think.

To start off, you're made of about 37 trillion cells. But you're also carrying the DNA of about 100 trillion bacteria. And the DNA of about 380 trillion viruses, many of which have taken control of the bacteria, while others are tossing their DNA into your nucleus, or even writing their DNA into your genome. Your cells need energy, so in each one of your cells, you've got thousands of mitochondria, and its 16,569 base pairs would not match yours–this DNA comes only from your mother's egg, with zip from papa. Also, half your cells use the genes on the X chromosome from your mom, and half use the genes on the X chromosome from your dad. Then you've got another 30 trillion cells with zero DNA in them: red blood cells. Coursing through your bloodstream and flying around your cells are short segments of RNA; for every one copy of your DNA, you've got thousands upon thousands more of it in short chunks of RNA, acting in all manner of ways–silencing genes, carrying messages, triggering molecular pathways, chemically altering longer RNA. If you're male, the 250 million sperm in you don't carry a "copy" of your DNA; they're each unique, and each a random scrabble of your DNA. Then your immune system has 4 trillion T cells and 10 billion B cells; most of them carry a unique amino acid sequence to manufacture a receptor for a foreign invader that the body has fought in its past.

So that huge diversity of genetic code is all at work in our bodies.

Critically, it's not just "the code" that matters. The three-dimensional shape DNA folds into matters (and it varies). Its location in the nucleus matters (and it varies). Chromatin is the cytoskeleton that wraps DNA up; over a lifetime, it gets tagged with chemical groups that alter where and when DNA is unzipped and read. A similar thing happens to the code itself; if you imagine your genetic readout as a nicely printed pristine laser copy of the Constitution, in reality it would have handwritten notes all over the place, lines crossed out, edits called for, sentences rewritten. And these edits vary greatly from cell to cell, or from one part of the body to the next. Another way to say this is that the genetic program you were born with is not the genetic program that's running in your cells today.

Then just look at the code itself. Forty-four percent of it are *transposons*, or, as they used to be called, "jumping genes." Yeah, it happens. Imagine if sentences in the U.S. Constitution just jumped around. Imagine trying to get an accurate tally of the books in a library if 44 percent of the bookshelves could wander off and get on elevators. It can be good for you or bad for you. If it's bad, your body tries to silence the jumped gene segment with piwi-RNA.

This is my personal opinion, but I'd argue that the most important component of your genetic machinery is not your code–it's your ability *to repair* your code. The exact letters of the code don't matter as much as your body's ability to fix it back, under the constant attack of life. It's under attack from the inside and outside. As Arvind wrote in the chapter about his mom's cancer, our DNA is mutating and creating cancer cells all the time, every day–but if things are working correctly, your body finds and kills them before they multiply into "cancer." Then our Earth is pounded every day by radiation, which rips our DNA apart. As our cells use oxygen, our body splits them in two, producing free radicals–these unpaired electrons go looking for an electron to pair with, ripping them off other molecules, including DNA. Which in turn rip an electron off the nucleotide next door, in a long chain reaction that has us, at all times, being ripped apart from within.

Another way to say this is, it doesn't really matter what kind of car you buy–what matters is that you have a good mechanic.

Now absolutely zero of what I've described is going to show up on a 23andMe report. Most of what's in a 23andMe report are "associations," often very soft statistical abstractions between cohorts of people. Too little of what it trawls for has any defined biology.

A venture capitalist could *never* fund a company that relied on an association study. They're not bankable. Conclusions should rarely be drawn from them. For instance, if you had all of the fifty-two genetic variants for higher "intelligence" (which I put in quotes because intelligence is a human construct, not a biological thing), you would *not* have an IQ of 180. You would have an IQ of 107, or seven points above average. That's what all those fifty-two genes add up to. Not much. But nobody has all the fifty-two variants. Almost everyone has a mix. So in most people, all these intelligence genes add and subtract against each other, to result in, on average, a single IQ point. *The intelligence genes do not determine your IQ.*

Even if a gene is "on," it might be highly active, or it might be minimally active, or it is conditional. A lot of the genome is conditional to the environment. It interacts with the environment. If it was computer code, it would not say, "Make Protein X." Instead, it would say, "If K is happening, make X." And another gene's program would say, "If K is *not* happening, make Y." And a third gene's program would read, "If gene1 gets lazy, activate the backup plan for making X out of Y." This is how genetics doesn't override the environment but responds to it. Genetics is not a fait accompli. Quite the opposite. It's through genetics that the environment shapes us so deeply. The environment is the ultimate gene editor.

Now there are important exceptions–but these are the Mendelian ones we all know, the ones that society first learned, which so influenced our determinist mental model for genetics. I have a niece with Down syndrome. Watching her grow up, it's undeniable that her third copy of chromosome 21 altered her life. We knew this the day she was born. In a similar way, anyone with single-gene rare diseases has had (until now) a predetermined fate.

Hemophilia is an example. One variant of the F8 gene causes a deficiency in blood clotting, and if it happens, the genome does not have a backup plan. It can be inherited or a random mutation. The scientist who sits next to me, Cody, is partnering with a gene therapy company to fix it. The interesting thing about gene therapy is it *doesn't* edit your genome. You might think it does, for why else do they call it gene therapy? But in the earlier days of gene therapy, they used viruses that actually *did* integrate new DNA into the genome. However, that proved to be very cancerous. So the solution was to use viruses that just injected the new DNA into the nucleus but not into the genome. Floating there in the nucleus, loose, it will still be found by genetic machinery, which will make blood-clotting proteins from it. The patient's genome is untouched and operates as always. The only problem–and the reason these gene therapy companies are coming to

Cody–is when cells divide, they don't copy and pass on the floating DNA. Cells are turning over all the time, and so right now gene therapy works for only a few years. By then, whatever you injected is all gone.

It's deeply ingrained into our society's moral framework that gene therapy is so dangerous *precisely because* it's a permanent change. Making permanent changes feels taboo in a way that temporary changes do not. Except it *isn't* permanent. Another scientist here today, Tim, uses a gene therapy pill to target epithelial cells in the intestines; those turn over in just four or five days, so quite literally, his artificial DNA is gone from the body in a week. (Unless you take another pill.)

It's really important for society to engage in these conversations around what's acceptable (or not) in genetics. Now that humankind has free will over the genetic system, all of humankind should speak up. But it's gotta be based on the reality of genetics, not the myth.

I suppose the Bourne movies are a kind of Rorschach test. You can see free will in them, or you can see determinism, and you can interpret them either way. What I choose to see in them is quite Nietzschean. Bourne shows that free will is possible–but only if he fights for it. Free will does not come easy. The safety of falling back into our cultural programming is all too tempting.

I hear that there may never be another Jason Bourne movie. But I'd like to see one. It was never made clear exactly what the Treadstone program did to Bourne to control him, so it'd be cool to find out the CIA edited his genome. They turned David Webb into Jason Bourne at the genetic level. They altered his pain receptors, they juiced his muscle reaction time, they tweaked his COMT gene so he could think calmly under pressure. They made him a tetrachromat, so he could see millions more colors. They gave him super endurance with a myostatin knockout. Bourne kinda likes all these things, but so great is the desire to be his true self, he seeks out rogue scientists to turn him back into David Webb. He has to return to his family home to find something with his original DNA on it–or maybe the CIA kept a copy–and Bourne breaks into their lab and steals it. But, the rogue scientist warns Bourne, he can make him David Webb again only temporarily. He doesn't have the same tech as the CIA. Bourne can be David Webb for only four days. He does it, and goes to see his mom. An assassin is waiting for him. Webb is caught, but he ultimately escapes in a way Bourne never could–by persuading the assassin that he too should escape his programming.

"You have a choice."

18

Iceland Bids Farewell to "OK Glacier," the First Glacier Lost to Climate Change CBS News

I am at Hellisheiði, the ON geothermal power station half an hour east of Reykjavik. This station powers most of Reykjavik and supplies electricity and hot water for the city's homes, buildings, streets, and stations, and because the geothermal power is efficient to capture, they pass along the low costs to residents. My in-laws' home, and every home I have visited, is kept at a balmy seventy degrees or more, even when it is below zero out. Because they can. The hot water is piped thirty-two kilometers from the geothermal site to the city, and then into water heaters in Reykjavik homes.

Hellisheiði is a marvel of geothermal engineering that plugs directly into Iceland's burning heart. The land of fire and ice got its name from the walls of steam bursting from cracks in its land. That steam comes from more than two kilometers underground, where active volcanoes heat it up. Where there isn't steam, there's glaciers, each of them nine thousand to twelve thousand years old. Iceland is the leading edge of a tectonic plate sticking out of the ocean. Iceland has 130 volcanoes–thirty of them active. (California has only seven.) Iceland is young. Its rocks are young. It's an arctic desert, a lava field, full of sharp volcanic features, and rocks so full of air they could float. Iceland has an otherworldly look that inspires Icelanders' belief in elves and magic to this day.

This special terrain–coupled with the unusual fact that electricity here is free–has created unusual conditions for incubating climate change technologies.

Standing in the windy cold and enveloped by billowing clouds of sulfuric rotten-egg steam, I am staring at a possible Band-Aid for climate change.

In a few shacks, they're running CO_2 capture technology. One is sucking CO_2 right out of the air. The other is taking the geothermal steam, which has ten times more CO_2 in it than air, and absorbing that CO_2 with a filter. My guide makes these CO_2-absorbing filters out to be some sort of magic, a marvel of chemistry–and other visitors are wowed. But I've seen CO_2-sucking technology before. I'm less impressed. Our lungs do the exact same thing in our capillaries, binding CO_2 to hemoglobin, ferrying it somewhere else, and releasing it. In this facility, their filter uses amines–which, when they're not performing the exotic task of grabbing CO_2, go by more common names, such as morphine, novocaine, or ephedra, the decongestant. Not the newest invention in the world. The only thing unique here is that in Iceland, they can run their fans on cheap renewable energy.

But once you've captured this CO_2, it gets really interesting. What are you going to do with it?

That's why this facility is special.

In a few places around the world, energy companies have been capturing CO_2 and pumping it into caves, then capping off the cave with the equivalent of a big wine cork. It's exciting, but the number of caves available to become underground CO_2 tanks is a real limiting factor. Others are pumping it into greenhouses, where the veggies grow 30 percent bigger. Though in that case, the plants release most of that carbon when they die. It's great for indoor crop farming, but not a climate solution. Some even compress it into CO_2 tanks and sell it to soda-bottling plants. But the world drinks only so much soda, and as soon as we drink it, it comes right back into the atmosphere. One burp at a time.

In Iceland, the very volcanoes that create the steam also created a rock called basalt, which is porous and spongelike in appearance. The possibility of filling it with CO_2 has been tempting. This too had been tried elsewhere, but the problem was the CO_2 came right back out. The sponge sucked it up and wrung it back out.

In Iceland, a kilometer beneath my feet, they don't just keep the CO_2 in the sponge—they turn it into stone. Rather than coming back up, it'll stay down there for more than a million years.

Ten percent of Earth's crust is basalt. Most of it is under the seas, and maybe we'll figure out that engineering someday, but there are plenty of locations on land available. Much bigger than Iceland. For instance the Pacific coast of the United States—from Alaska to California—is both volcanic and basalt.

At this facility, the pumping stations are geodesic igloos. They look like moon bases.

The alchemy for turning gas into rock looks like this:

$$(Ca^{2+}, Mg^{2+}) + CO_2 + H_2O = (Ca, Mg)CO_3 + 2H^+$$

"We thought it would take fifty years to turn into stone," my guide says. "What was so amazing was that it only took two years."

What made that possible was the unexpected benefit of the mineral olivine. If you don't know its color already, you can guess. There's a massive amount of olivine underneath Earth's crust, but in Iceland, enough of it found its way up into the crust to play a critical role in the chemistry needed to speed up rock formation. Just as a bonus, as the rock forms, the water pumped down with the CO_2 precipitates out as purified groundwater, and can be drawn back to the surface as drinking water.

I'm pretty excited to see this—to be here—even though I'm not actually seeing it. Just standing on this part of the land, knowing what's going on underneath, has me pumped. Then I ask the guide how much CO_2 they've sequestered so far.

"I think it's twelve thousand tons," he says.

I pause. I think. I do some math in my head.

I'm no longer as pumped.

Every year, humans on Earth generate 37 billion tons of CO_2. Even Iceland, which barely has any emissions at all, generates over 5 million tons of CO_2. Just to compensate for Iceland's greenhouse gases, they'd need to create almost five hundred of these plants.

Driving back to Reykjavik, I realize that for almost all of us, climate change is not really something we live with every day. I mean, in America and Europe we've got storms and heat waves, but we always had storms and heat waves; today's are just stronger. Most of climate change is in the future or only on the news. In the United States, there are only so many people living with actual climate impact. Drought in the piñon trees in New Mexico, tidal flooding in South Carolina, some Cajuns losing their land in Louisiana.

But in Iceland, it's right out at the edge of town, where entire glaciers are melting. Ok (pronounced "oak") jökull (pronounced "yokul") had the glacier title stripped down a few years ago to just OK and became known as the OK Glacier. But it wasn't OK. It melted this year, and just two weeks before my visit, the Icelandic prime minister gave it a proper burial, with a tombstone plaque that reads:

> Ok is the first Icelandic glacier to lose its status as a glacier. In the next 200 years all our glaciers are expected to follow the same path. This monument is to acknowledge that we know what is happening and what needs to be done. Only you know if we did it.

Icelanders are stubborn and independent. They are shaped by the stark landscape–their music is massive in sound (Jónsi, Björk) and harsh and experimental. Yet all Icelandic mothers still knit wool clothing. Though tiny in population, they are on the world's stage, making headlines for their economic booms and busts, their politics, and their soccer team's postgame

Viking clap. They don't care what others think, which liberates them, and is the key to why they are avant-garde. Having families together without marrying is more common in Iceland than anywhere else in the world–and they take pride in skipping the shackles of that convention. All children are taught to sing and play music.

Fourteen years ago, with my now wife, I visited Jökulsárlón, the iceberg lagoon, a main tourist stop in the southeast of Iceland. I was stunned by a steel gray bay full of floating palaces. Beautiful and bizarre icebergs in a dazzling variety of blue and white hues. Twisting towers of turquoise to cerulean to ultra white and back again. As if the entire rainbow danced between blue and white, sometimes streaked through one 'berg. Seals lay lazily on the rafts of ice, hitching a ride. Occasionally, two house-sized chunks would float into each other, like bumper cars, flipping one upside down, with a roar and rush of seawater raining down, creating a misty rainbow. Kristrun and I sat on the moraine and watched the geologic show for hours.

Ten years later–just four years ago–I visited again, to the same spot. I blinked into the reflected sun glimmering on the bay's surface. Devoid of the massive ice, only much smaller chunks of relative ice cubes were left for me to see. The difference in just one decade was a shock. Jökulsárlón's stark splendor had died to me.

As I drive into Reykjavik, I stop by Laugardalur Park. There was some tree planting here done by horticulturists, and I had to see it for myself to figure out if it's art or if it's forestry. Because these trees certainly make a statement.

They're palm trees.

Yes, they're planting palm trees in Reykjavik. They've been in the ground just a few weeks. They're not the soaring royal palms of California and Florida. They're actually imported from the Himalayas, and supposedly they grow fast. The city seems dead serious about seeing if they'll survive the winter. Though when I look at them–there's five of them, with their stubby trunks and green fans–I still can't answer my own question: if they're a wry comment on what Iceland's becoming, or just a practical solution with unintended ironic effect.

Wild, pink-footed goose season has just started. My brother-in-law had gone hunting, bagging one and roasting it. I stuck to the potatoes, but my family picked pellet shot out of the meat as they ate. Meanwhile, they pick my brain only a little at dinner; their curiosity about carbon sequestration lasts barely a minute. Even my fourteen-year-old niece, who considers herself a climate activist. They know climate change is real, but they actually don't understand it any more than the rest of us really understand how our mobile phones work.

Today, Earth has become the coal mine, and Iceland is the canary.

A miniature version of the entire carbon cycle plays out here. And what makes Iceland's carbon cycle so lucid is that it doesn't just run its course in geological time, over hundreds of millions of years, but also in real time, in decadal sprints. In the eastern United States, a geologist might point out, "Four billion years ago, these hills were volcanic, and ejected carbon into the atmosphere." In Iceland, teenagers will say, "Oh yeah, when I was a kid, Eyjafjallajökull was really blowing."

For a month of 2010, Eyjafjallajökull spewed 150,000 tons of CO_2 per day, wrecking transatlantic air travel. As rain fell–as it did that season four days out of five–the CO_2 would dissolve, turning into carbonic acid. A weak acid rain, pelting the Kálfafell plains and the Vatnajökull glacier. The following year, Grímsvötn erupted, right beneath Vatnajökull, exposing fresh new rock that's particularly susceptible to the acid rain, where the CO_2 strips the rock of calcium, sodium, and magnesium, washing right down Tungulækur river into the North Atlantic basin, or taken east by the rivers to the salmon farms on the coastal fjords. On the cold-water shallow shelves grow blue mussels, scallops, a strange tubular mollusk called tusk shells, and tons of phytoplankton, all of which find the free-floating CO_2 and calcium ions to make calcium carbonate to build their shells. When they die, they sink to the trenches of the Atlantic and the Norwegian Sea, where they sink into the earth. Calcium carbonate is slowly heated and pressured back into its elemental parts, becoming magma again, awaiting the next eruption.

Humans emit more CO_2 each year than all the world's volcanoes do. Not a little more, and not twice as much–120 times as much. What goes up today does not come down. We're going to become like Venus, which baked out all the CO_2 in its rocks, then got even hotter. NASA once sent probes to Venus–the probes lasted only a couple hours before succumbing to the heat.

At IndieBio, we don't deliberate at all on the moral question of whether climate engineering technology is needed. Many environmentalists are opposed to climate engineering, simply because it gives polluters an out. It's far better, they argue, to simply stop burning coal and gas. And we don't disagree that would be ideal–we just disagree there's any realistic chance of that happening. Even in Iceland. Inertia is too entrenched. As well, there are certain industrial processes, such as smelting metal from ore, where electric furnaces can't get hot enough. (Iceland's biggest source of emissions is smelting aluminum.) So remediation technologies can play a critical part–at least in the next decades, and likely for the next millennia.

We might dream of a future with no smokestacks, but that future isn't in the next twenty-five years. Today, smokestack technology is so good that it can capture 80 to 90 percent of the CO_2. The United States gives a $35-per-ton credit for capturing carbon dioxide. But still, the question is what to do with it, and there's no point capturing it if there's no place to put it. Most don't.

In the United States, the number one biggest user of the tax credit is the Petra Nova facility near Houston, Texas, which sends the CO_2 from the coal boiler eighty-two miles away, where it's used to frack yet more gas from an oil field. Oil production in that field has gone up fifty times. So our carbon credit is generating *more* oil extraction, not less.

It's a wicked game, where wolves hide in sheep's clothing.

CRI is a company in Iceland with a better idea, and so I sit down with their CEO, Inglo, and Margaret, their head of business development. This is in Kópavogur, a suburb of Reykjavik, at CRI's headquarters.

CRI makes the same things that oil companies do–gas, plastic, resins, and paint–but without extracting any oil. They capture the CO_2 from geothermal steam and have found a clean way to convert it into carbon-neutral methanol, which is sort of a universal Lego block for biofuels and bioplastics. CRI is about a decade old, and all over the North Atlantic their fuels are used. Swedish cruise ships, shipping tankers, ferries, even small pilot boats. A lot of these vessels proudly display a green flag, so others know. It costs a little more, but everyone is willing to pay.

"There's a green tsunami surging through the value chain," says Inglo.

Margaret explains how they're expanding to Germany, Norway, and China.

I ask about the United States. I try to explain that the green tsunami will soon be here. But we are not in their plans.

With Icelanders, there's very little subtext when they converse. They say what they mean and they mean what they say. They know they can't save their country all on their own. No matter how much CO_2 Iceland converts into rock, their glaciers and fisheries are dependent on what happens in China, the United States, and the other big economies. But they're not complainers. Or let me put that more succinctly. They're not whiners. If Europe's not doing its part, they'll be sure to say it. But without whining. Iceland manages this psychological predicament–that their fate is in the hands of others–by staying true to the root of their culture. They're independent people. Doing their part is a big deal to them. Doing their part is precisely how they lay claim to independence.

I take these thoughts with me into my next meeting, which is at the Ministry for the Environment and Natural Resources. Helga Barðadóttir is a head of division here.

Helga takes me through Iceland's entire climate plan through 2030. It's a lot of detail. They are building charging stations for electric vehicles everywhere, and don't just tax the aluminum smelters but cap their output. My brother-in-law used to work at one; they have become very controversial. The Green Party can push only so far; if they're voted out, good work would be undone.

Helga explains the government is very worried about the Gulf Stream moving south, and taking the fish with it. They're even more worried about sustaining tourism–people visit from all over the world to see the lunar landscapes topped by giant ice sundaes, but they're melting like soft serve. Trees don't grow so great here, but reforestation is being tried. Twenty years ago, my uncle planted hundreds of trees on his summer house property, an hour away. They are still spry and thin.

What Helga really wants me to know is, *We are doing our part. We have a plan.* But then she speaks with a barely detectable frustration. "Climate change is a global problem and we cannot solve it ourselves." That's as much subtext as you'll ever get out of an Icelander.

As I leave the meeting there's a light rain. I walk into Laugavegur, the trendiest shopping area in the trendiest country on Earth, and my mind turns. I'm surrounded by these glass showrooms, displaying steez interior designs and effortlessly attractive apparel, and I start thinking about all of Iceland as a showroom. A showroom to the rest of the planet.

The only way Iceland can succeed is if it manages to inspire tons of imitators–to start a huge trend. But I'm going to make an unusual argument here. Earth's atmospheric climate is inextricably linked to its political climate. It requires political consensus to alter the economic incentives. So Iceland's CO_2 technology is only one aspect to display in its showroom. The other aspect is its political climate.

Iceland, even more than the rest of the world, was an economic wreck after the debt crisis in 2008. Most of the last decade, it was chaotic. They devoted four years to rewriting their constitution, only to then abandon the effort. Protesters threw eggs and shoes. Iceland had to have four national elections in eight years, as political coalitions kept falling apart. Political discourse was insulting and abusive, not just online. New political parties kept cropping up: the Pirates, the Dawns, the Right-Greens. Even when parties stayed in power, it required turnover at the top–they had four prime ministers in just two years.

All of a sudden, they stabilized. A Green Party academic, mother of three–a former protester–forged a coalition with two alpha-male old-guard neoconservatives. The Greens got to run environmental and health care ministries, while the NeoCons got to run banking and fiscal policy. It's an unlikely fusion; they have little in common, honestly. But I think the twin disasters of climate catastrophe and financial catastrophe changed the chemistry and forged this unusual bond; it took near ruin to force change on the system. They stopped being idealistic and became pragmatic.

We have as much to learn from Iceland's political compromise as we do from its carbon technology.

Back at my brother-in-law's house, I am hanging out with my niece, Isabella.

She is fourteen and, like all kids her age, not worried about boys but about climate change. She shows me her bedroom wall. When I was fourteen I had sports posters and rock bands all over my room. Isabella has a map of Iceland with newspaper clippings chronicling the creeping death of climate change.

She's raising a baby parrot. Next to the map is a small notebook paper, torn from its spiral binder, with pen scrawled in neat writing. Its words eerily echo the prime minister's plaque at OK Glacier.

> My bird's name is Alda. If we are not here anymore from climate change, please feed her and give her water twice a day. Or she will die. Thanks.

Pathos is an easy effect to trigger, but Isabella isn't going for it. Her eyes steel. She wasn't feigning concern; she *was* concerned. I think she might have been dead serious when she wrote it.

Isabella says she recycles every scrap and has got the family to eat less meat. They don't order off Amazon, because of the shipping distances the product travels. They buy local, despite the cost. She even says she's careful with pencils. Rarely sharpens hers. Only when she really has to. For the trees. Telling me this brings a slight smile to her face. Like she's told me a secret.

And then I get it. Being a young activist may not save her land, but it saves her psyche. She feels empowered, instead of helpless.

Doing her part is the Icelandic way.

19

Playing God: "We Are in the Midst of a Genetic Revolution" CBS News

You would never meet one of our geneticists at an event and then walk away thinking, "Man, that guy reminds me of God."

So we hate the broadside characterization that geneticists are "playing God" when they engineer a genome. It's a tone-deaf phrase, failing both as an analogy and as a declarative.

When geneticists step out of our lab, they don't act like people who just had a holy moment. If they've had some success, they're excited and they feel lucky, because they know what they do is really hard to figure out and get right. They don't feel touched, and they certainly don't begin every bench session with a spiritual ceremony. They're not looking for a flock of disciples.

To a geneticist, altering life is not in the same league as creating life. Altering life is just, well, like what evolution and natural selection do. When a wolf attacks a sheep on a ranch, the rancher doesn't stop and say, "Man, that wolf is *playing God*." No way.

But creating life—out of lifeless components—*if* it could be done—now *that's* playing God. If a scientist could re-create the origin of life, turning primordial soup into some sort of protobacteria like Earth first had 4 billion years ago, then they would indeed be heralded as a demigod.

Hanging out at IndieBio after an event, late at night, can feel like a college dorm all over again. And inevitably, at least once every new batch of companies, the conversation strays to the origin of life.

Once you crack open that topic, within a few minutes you'll have a half-dozen physicists and biochemists and computer scientists at your desk, wanting to contemplate this Holy Grail. Usually, one or two of them know a whole lot more than the others about molecular thermodynamics and self-replicating RNA, and they start to hold court a little bit. But the others can follow along perfectly well. It's a fairly giddy moment to find coconspirators.

We had one of these moments yesterday. It just...happened. A few scientists overheard us chatting—and then, like spontaneous combustion, a bullshit session on the origin of life broke out.

But this session was different. Because Shane Kilpatrick had brought his notebook of equations.

Shane is the CEO of a startup called Membio. He worked in bioremediation in the tar sands of Canada, where he learned to capture methane pollution and transform it into more desired compounds. He brought that biochemistry to the realm of human health care and started a company that makes synthetic blood—or, more specifically, that makes red blood cells from scratch. In that sense, he's sorta growing life out of a bioreactor and some raw ingredients. But red blood cells are in the gray area between alive and dead. Technically, they live for 120 days, then die. But even while alive, they don't really meet one of the most important standard scientific definitions of life: red blood cells can't self-replicate. Red blood cells are more alive than fingernails, but not as alive as taste buds.

So Shane had a leg up when it came to speculating on the origin of life. And he admitted that, one Saturday, not too long ago, he'd worked out some of the math. (He had nothing better to do that day.) He had filled seven pages of graph paper with fine print equations.

Now just to be clear, anyone involved in these bullshit sessions implicitly understands the rules. Any one of us could go read the papers from the top three or four researchers in this field, but that kinda feels like cheating. It's more fun to see if you can figure it out by yourself. Or at least to see how far you can get before you have the overwhelming urge to go look up the answers. And if you do it yourself, there's always the scant possibility you find an original angle.

As we went through Shane's equations, it became clear he had a lot of the possible chemistry, but it was lacking unifying theory. Just such a theory emerged in our conversation, and we took it one step further–proposing a pretty simple experiment we could do in the lab.

We weren't trying to create life per se. Rather, we were trying to create something "more lifelike" than what we started with, something resembling an extremely primitive protocell.

Starting Point: Not-Life Soup	**Goal: More "Life-Like"**
RNA RNA polymerase (an enzyme) Loose nucleotides Loose sugar phosphates Fatty acids (fats) Fatty acid synthase (an enzyme) Water/Buffer *Nothing alive here!*	By varying the environment for this soup, we hope the fats will self-organize into a membrane, accidentally trapping the RNA inside. The RNA will self-replicate, and fatty acids will form, resulting in a stronger/thicker membrane.

Don't get excited. Ten years from now, this experiment would barely win an eighth-grade science fair.

After we set up our experiment and had it running, only then did we look at the research from the leading scientists in the field. We were definitely in the ballpark. Our goal for this chapter is to walk you through that ballpark, at a conceptual level, but to do it very slowly, so that ideally you have the epiphany *before* we've finished spelling it all out. It resembles an optical illusion, or putting puzzle pieces together, in the sense that all of a sudden the mind intuits something previously unseen.

Imagine a filmstrip that captures the origin of life. Over the series of frames, it begins with parts needed for life, but what's in that frame is clearly not yet "alive." Near the middle of the filmstrip, a living bacteria has formed. Bacterium1. Keep the film going a bit, because if the bacterium falls apart a second later, then it doesn't count as life. It's life when Bacterium1 *stays* alive.

Why this *seems* so improbable–and why it's so fun to contemplate–is that the origin of life seems to defy physics. The second law of thermodynamics says that entropy always increases. A system should always have less and less available energy over time. A system becomes more disordered over time, and more unpredictable, more random. All things decay. The laws of thermodynamics allow no exceptions. But life seems to be just such an exception. Life seems patently *more* ordered than nonlife; and that order means life-forms are quite predictable, and the higher the life-form, the more organized, the less random. Life-forms also seem to gain in energy, not decrease. Life-forms grow rather than decay.

But life does not defy physics–not one bit. So there's something misunderstood in the eyes of the observer here (us). An illusion hiding in biology.

Let's lay down some clues.

One clue is that at the beginning of time, some 13.7 billion years ago, all the universe's energy was infinitely compacted. One way to construe the big bang is that it dispersed that energy, which continues today, toward infinite dispersal. Over time, energy will always get more dispersed. The sun will disperse all its energy; a rock will eventually decay into dust. You might say the universe prefers better dispersers (but this is something we'll work into). Life fits into the picture if we can see biological life as *more dispersive* of energy than nonlife.

Now let's talk snowflakes.

When they land upon the ground, snowflakes appear exceedingly ordered, intricate, complex, and symmetric. There is no life in them. But if you compared a snowflake to hail–a mere frozen raindrop–you might say that the snowflake looks a little more lifelike than the frozen raindrop, exactly because it's so complex. The energy source that starts a snowflake's first crystal is the thermal gradient: a sudden drop in temperature. As a snowflake builds from that initial ice crystal to its kaleidoscopic lace, at each step it releases energy as it gains another crystal. A snowflake is an example of order increasing from energy dispersing.

Something snowflake-like happens when you put a magnet to iron filings. Rather than clumping in a ball around the magnet, the filings align into radiating projections, like the swords of the Iron Throne. The filings organize instantly, as their electrons align to the flow of the magnetic field. If you took a photo in that moment, it might look like you had captured an image of an explosion–when it's actually a kind of explosion in reverse.

So there's this phenomenon of lifeless materials increasing in order and complexity in the presence of an energy source. As rocks are melted in the core of the earth and that magma rises to the surface, if it cools slowly enough, it will vibrate, release energy, and re-form into the crystal quartz. Salt crystals form as salt water evaporates. Under intense heat, ordinary sand reorganizes its atomic structure into glass. You could think of these more orderly structures as the material equivalent of harmonic musical notes; a sliver of reed vibrates, making almost no sound at all, but the vibrations are picked up by the air and the body of the instrument, amplifying it, releasing more energy to make more sound.

In the earlier chapter at the Caltech observatory, where astronomers are looking for signs of life on other planets, we explained how young stars shoot out amino acids and nucleotides–such that every planet has likely been littered with the building blocks of life. For life to start on Earth (or even if it started on another planet then got sprinkled here), it could use these intergalactic building blocks, orderly units forged in the nuclear fusion of young suns.

We are a rocky planet, so it's appropriate to compare rocks to man. Let's imagine a 180-pound rock versus a 180-pound man. And let's think about the energy they disperse over time. A rock has a decent amount of energy contained in it–mostly chemical energy from all those rock-hard bonds. But over one hundred years, a rock barely disperses any of its trapped energy. During the days, it absorbs the sun's rays, heats up, and then emits that heat slowly over the rest of the day. It decays slowly, with acid rain. In a hundred years, it's not made any dent in its ecosystem (unless it rolled down a hill and crashed upon some things). It'll take ten million years to turn that rock into dust and release its energy.

But a 180-pound man–the universe could not design a better trafficker of energy. Just like the rock, he soaks up the sun and disperses this heat with his sweat and breath. But he also eats carbon sources all day long, digesting it into energy, with which he runs around, makes war, and makes love. He even grows carbon sources in his fields, and with all this food around, he replicates in plurality. He makes more humans. They chop down trees to burn. They dig up the terra firma and find black rocks to burn. Man builds machines that in turn harvest and dissipate yet more energy. In the same hundred years that a rock does jack squat, a man disperses the energy of a mountain.

To the humble stored-energy sources in Earth, man is chaos. Dispersing it far and wide. Man pleases the entropic laws of the universe.

By this logic, you can begin to see that life-forms, and order (or complexity), can align with and obey the laws of the universe. It starts to seem conceptually possible. Earth has constant sources of energy (the sun and the molten metal core at the center of Earth). That energy can, by itself, reorganize inorganic matter into more and more ordered forms, each better at releasing energy than the previous form. But how the first cell forms is still a long way away.

Cells are self-replicating. Life is self-replicating. Bacterium1 divides in half, making Bacterium2. But even though self-replication is a necessary criterion for life, self-replication happens even in not-life circumstances.

The simplest example of self-replication happening is fire.

Fire is technically lots of nanofires, a self-replicating chain reaction of them. With the initial strike of a match, or a bolt of lightning, a molecule is pushed past its flash point and ignites. That releases energy, but so much energy is released that it overheats the molecule next door past its flash point, which ignites in turn. It keeps going until it runs out of fuel.

The fire in our bellies operates in much the same way, but it's more of a chemical ignition. Glycolysis turns a molecule of sugar into two ATPs. One is harvested by the body for energy, while the other is used to ignite the next sugar molecule. Our bodies have evolved to harness this self-replicating ignition.

This is a good time to break the narrative into stereoperspectivity.

It's clear that when Po and I break our collective voice in two, we bring more chaos to the page.

Ahh–nice one. By dividing in two, we release more energy. Very meta.

It's pretty cool we haven't overwhelmed the reader with technical detail yet. Snowflakes, magnets, fire. Rocks.

But now we've got to explain RNA. And enzymes.

We'll go easy on them.
RNA is a little intimidating, I know.

Yeah, the appearance of RNA in this story makes the reader worry, "Uh-oh, they're about to lose me in physics and biochemistry!!"

But you can't understand the origin of life without RNA.

If you looked at RNA, without knowing it was RNA, it wouldn't "wow" you. It's just a string of sugar molecules with some nucleic acids attached.

But the key thing about RNA is that it's self-replicating. Much like fire.

Okay, let's go back into mono mode and explain that.

The extremely cool thing about RNA is that because it can be strings of any length, it can work both as genetic code *and* as an enzyme. A shorter string of RNA can float near a longer string, chemically bind, and trigger a chemical reaction that causes the longer string to replicate (assuming the materials needed to build a copy are nearby—more sugars and nucleic acids). This may seem a heavy load for a mere enzyme, but enzymes are the magic pixie dust of the biochemical world. Enzymes are the bridge between chemistry and biology. Enzymes are not given their due credit for being central to making life run. So often we ignore little enzymes in favor of studying and analyzing their big brothers, proteins.

What enzymes do is offer an alternative chemical pathway—one that takes radically less energy to pull off. There are a dozen or so chemical reactions—all very common in biology—that would take a *lot* of energy to trigger without the presence of an enzyme. You can sorta think of an enzyme as prebuilt to facilitate certain chemistry. And in this sense, you can think of biology as a more efficient form of chemistry. Another way to say that is that biology can do energy dispersion with less energy input than chemistry alone. Just a few sunrays can keep life going.

So the origin of life was not the start of things self-replicating. RNA was already self-replicating, and life borrowed it.

Another rule of life is that all life-forms have a membrane. But spontaneous creations of membranes are the easy part, and maybe the most well-understood piece of the puzzle. All it takes is some fat molecules and an appropriate source of energy, such as a pH gradient. Just as oils will congeal together in water, there's an advanced version of this. Fatty acids will absorb energy and then self-organize into supermolecular structures called micelles. Taking this intricate, spheroid-snowflake shape allows it to release more energy than it took in. In turn, micelles can also absorb energy and then transform into ovoid membranes to again release yet more energy than they took in. A membrane is the bioequivalent of quartz.

So all it takes to get a very primitive protobacteria going is to induce a membrane, accidentally swallowing (or surrounding) some self-replicating RNA.

And that's the concept behind the experiment we are doing in the lab. Inducing a membrane to form around enough of the right genetic soup to get the biochemistry going.

But let's flip ahead in the filmscript to when it did happen in nature, some 4 billion years ago, in the presence of just the right energy sources. It's not going to last–it's not going to *survive*–unless it proves itself as a superior energy disperser. The sum has to be better than the parts. So as this protobacteria absorbs energy, from heat or pH or whatever, it needs to continue releasing more energy than its parts would do on their own. It can do this any number of ways. It could eject a molecule now and then. (Eventually, it will eject a whole copy of itself.) It could instead hang on to that molecule and make it part of the membrane, releasing chemical energy as it slides into place. (Eventually, it will learn to build fantastic structures.)

So that's the basic concept of the origin of life. It isn't an exception to the laws of physics.

Hopefully, at some point while you were reading, you began to see the other side of this perceptual illusion. At every step, energy is released and entropy increases, but very orderly-looking things are the result. Life doesn't look like "decay," because we had an incorrect presumption of what decay looks like.

We're going to break into stereoperspectivity to discuss what this means for how we understand life.

I think we have to deal with the implication we
created that man is the ultimate chaos machine.

Yeah, let's revisit that.
I'm sure in many readers' minds, in the
moment when they read those words, visions of
human-made chaos flashed before their eyes.
As if we were arguing such mayhem is inevitable.

Even more than inevitable.
Actually the true or pure way to live.
Aligned with the universe.

But we don't mean that.

Correct.

In fact that's making the
same mistake all over again.

Absolutely.

The lesson should be: It's not through chaos that man disperses the most energy. It's through order. Complex order can be a lot better at fulfilling the destiny of the universe than chaos is.

Man doesn't release a lot of energy just shouting and banging rocks. Even if it seems like he does.

So my intuition is leading me to an idea. But tell me if I'm wack.

Go for it.

It makes me wonder if mutations of genetic code weren't the only thing driving evolution.

Okay.

It's like the universe selects for "fitter biochemistry." Or rewards it. It selects for any improvement in energy dissipation. And so biology got more and more ordered and sophisticated, simply following the laws of physics.

Okay, you're not wack. This is a theory
that science has really been thinking about.
Another way to say it is that evolution began
before life started. It was an evolution
towards more and more energy dispersion.

Woah...

Yeah.

If life is inevitable...

We didn't quite say that.
Inevitable under certain conditions.

Is it?

Maybe here on Earth. Every speck of
Earth, from its upper atmosphere to the
bowels of the inner earth, has been found to
have life there. 300-million-year-old coal
has been found to have bacteria spores in
it, which can be revived, just with water.

This is going to be a hard pill to swallow.

How so?

Is it nihilist? Reductionist?
It kinda means that we're not special.
We're not an exception.
We're not a one-time event.
We're not a miracle.

Even if biological life is inevitable–
our existence was not inevitable.
That we exist, with the lives we have–
we are as improbable and just as
much a miracle as ever.

Let's go back into mono to get to the end.

Nobody yet knows the *exact* way life happened; and to get from a proto-bacteria into a self-replicating bacteria would take tons of further steps. But a number of amazing scientists have reverse engineered steps of it, be that on paper, or in a computer simulation, or even in God-like physical experiments, where they put elemental building blocks inside a tiny gold box and heat it up to outrageous temperatures. Describing these varied efforts in detail is beyond the scope of this chapter. There are good reasons to believe the origin of life was more probable in a geothermal vent at the bottom of the ocean than in a nebula in space–because of the chemical and mineral ingredients available and the chemistry shortcuts that enzymes would enable.

We do believe that in the next twenty-five years, we'll see a scientist re-create life in a black box (or gold box). And this person—or more likely, team of people—will be hailed as truly playing God.

There are many other amazing efforts that will be described as "artificial life." Slowly replacing every nucleotide in a bacteria to make something with 100 percent synthetic DNA. Growing stem cells into artificial organs. Bringing species back from extinction. Inventing new amino acids and designing genetic machinery to use them. Deploying Hachimoji DNA, which has eight nucleotides rather than four. All of these will inevitably be characterized as artificial life, but they're really just super-whizzy ways to *alter* life into synthetic forms.

When life is finally created from nonlife, it's probable—even likely—that it won't be the same as life today. It'll probably be slightly different chemicals and molecules that come together. It will be new life. And that will be exhilarating.

My philosophy is action

"I've eaten dinosaur."

The problem is that we can't agree on the problem

The future we get will be the future we fought for

Fight capitalism with better capitalism

Whoever has the best biology wins

Nature is the greatest bioterrorist

Don't judge blue by its color

All technological revolutions are social revolutions

Dystopias require no imagination

The mute give voice to others

Inertia is the immovable object

This time, the meteor has a steering wheel

Bankers are far more dangerous than robots

Mankind's greatest illusion is money

Only with freedom to fail can you succeed

99% of censorship is self-censorship

Humans are the cheapest robots money can buy

If everyone agrees with you, it's not a future trend.
It's a present trend

Nobody is an expert in the unknown

There is no plan, just a way

The future of our food is the future of our species

The environment is the ultimate gene editor

Earth has become the coal mine, and ice is the canary

Biological life was inevitable, but our existence was not

Civilization narrows the breadth of human experience

Freedom is being a master of your own time

Aging is just a failure to repair

There is no such thing as natural

What life means is not decided for us, it's decided by us

Fertility is about equality

We're hopeless until we can measure it

AI isn't evil. Are you?

The future of editing the genome is not editing the genome

We still live off the land

The search for truth is the greatest adventure

My soul is where I fly

20

A Fitness Guru Who Goes by "Iceman" Says Exposure to Extreme Temperature Is a Lifesaving Third Pillar of Physical Health

Business Insider

I'm in Venice, Italy. The island of San Giorgio Maggiore. August 9, 2019.

Lie back in grass.
Blue sky through tree branches.
Breathe deep, fill lungs then throat then chest.
Exhale.
Over and over.
First buzzing began in fingers.
Then buzzing in diaphragm.
Then buzzing in forearms.
Then fear.

What will happen if I keep going hard?
Then hold my breath.
Buzzing turned into oxygen.
I can feel the buzzing become breath for my body.
Minute forty-five seconds. Never held breath longer than a minute.
Then breathe.
The birds and crickets got very loud all of a sudden.
The sky was vivid blue then the color drained out.
The branches waved above me in the wind.
Second round.
More comfortable now, push harder.
Bigger buzzing. Birds were loud. Branches waving. Neck getting stiff.
Two-minute breath hold.
Third round.
Ants crawling on me. Who cares.
Fourth round.
This is crazy.
Fifth round.
Timed the breath hold myself.
Two minutes easy.
Okay done.
Something happened, not sure what.
One guy is screaming yes, yes, yes.
Profound experiences all around me.
Buzzing lasted fifteen minutes.
Not sure the mechanism of action.
All we did was breathe.
A kiddie pool filled with ice cubes and water.
Need to last two minutes.
Freezing to touch.
Get in.
Shock.
Hyperventilating.

Get it under control.
Breathe. Breathe.
Inhale through nose.
Out through mouth.
Try to slow it.
Calm.
Last thirty seconds.
I won't die from this was my mantra.
Stay under control.
Two minutes.
Made it.
Did not feel warm at all, just pain.
Just warm up.

What you just read were my notes from the Wim Hof experience with Wim Hof himself. We spent two days together, with mutual friends, and on the second day he led us all through his routine. Wim is becoming famous; he sells out performing arts centers as he travels around the world, spreading his message.

Most people first learned about Wim for setting obscure world records–all of which had to do with extreme cold exposure. He ran a marathon in Finland when it was thirty degrees below freezing–in shorts. He sat neck deep in ice for one hour, fifty-two minutes. He got to Death Zone altitude on Everest–only in shorts. But over time, as he kept setting new records, people took interest in his training methods, and then in him.

When Wim and his wife were young, she committed suicide, leaving Wim with four children to raise. In the slow recovery from that, he developed a message for society, which is about learning to revive feeling itself, your aliveness. Wim found that aliveness in frozen lakes and the bone-cold canals of Amsterdam. He talks about reawakening our primal brain. Retaking the inherent power of our minds.

A friend of mine swore to me over dinner in San Francisco earlier in the year that the Wim Hof Method had cured his allergies. He takes cold showers and swims in the bay. Wim Hof followers don't stop there. They have claimed all sorts of benefits. More energy. Better sleep. Higher concentration. No hangovers. A replacement for pain medication. Relief for fibromyalgia. The list goes on.

I was excited to spend time with Wim, but I was also a little queasy. I felt awkward. And the reason was, pseudoscience makes me uncomfortable. It just does. I was worried that Wim might be selling metaphysical bullshit and invoking science inaccurately. I didn't want to play the role of the "science police."

Every day, we take about twenty-two thousand breaths. Most of the time, we don't think about it. It's easy to take for granted. It's unconscious, subconscious, automatic, autonomic. Like many things in life, it's taken care of for us, and we can devote our attention elsewhere. But breath is one of the few things in the autonomic nervous system that is simultaneously wired into the somatic nervous system–the system under our control. This dual wiring exists because there are times animals like us *need* to be able to control our breath. For instance, to cough to clear an obstructed airway. Or to sniff, to track other animals. Or to hold our breath–like a dolphin under the sea.

Breath, then, is a bridge between two realms. Most of the time, breath reacts automatically to what's happening in our bodies. Wim and other breath instructors propose that we can reverse that mechanism. By controlling the breath, you can control what's happening in your body.

The general idea of that is probably incontrovertible. The problems start when anything specific is mentioned. Can you really control your immune cell count? Can you change the thickness of the cortex in certain brain regions? Can you alter which hormones are released by the endocrine system?

These thoughts were in my mind as I lay down in the grass of the beautifully manicured lawn, staring at the sky through spindly tree branches. Wim booms, "Fully in! Let it out!" Then, "Fully in! At the top of the breath, hold it!" I do as asked, holding my breath for the third time. I close my eyes.

I become an oxygen atom sucked through my nose, swerving in laminar lines around nasal-hair tree trunks into the trachea and down into my left lung where immediately I hit an alveolus, the bulblike cul-de-sac. After a split moment I am flung by an attractive negative electric force through the cell wall of the sac and through the cell wall of a capillary and again through the cell wall of a red blood cell where I clang to a dead stop on one end of a hemoglobin molecule shaped like a four-way teeter-totter. Another clang! A CO_2 is ripped off another arm of the same hemoglobin, and the negative charge draws another O_2 in to replace it. Two more times in less than a heartbeat. The hemoglobin is fully loaded with O_2, and so are the other millions of hemoglobin molecules within the same red blood cell. My ride gleams bright red. Lub-dub. Off we shoot in the next heartbeat. I am staying in my red blood cell longer than usual. Normally there is plenty of CO_2 that wants to trade places with me on my hemoglobin ride. Not this time. There is much more oxygen in my body from my fast, deep breathing. I keep moving and moving and moving. A full two minutes later I finally get pulled off my hemoglobin ride into a cell, high-fiving the CO_2 replacing me on my way. I whiz through the cell membrane and into the mitochondria. My negative outer shell is hungry for electrons. This cell is relaxed. It wants me. NADH throws me an electron, becoming NAD+ and burning in the process. I am torn in half. My half body hungry for positive charges. A free radical. Two positive hydrogen protons find me and we link up, becoming H_2O. I have burned and am resurrected as water.

Humans are electric beings. We generate current and use it, just like a robot or a Tesla. We rip electrons off glucose and send them along a transport chain to oxygen. Oxygen is the most important molecule in our bodies. It disperses those electrons. Without oxygen, we would essentially slowly electrocute ourselves.

In the autonomic nervous system, there are two branches, two opposing systems that coregulate each other. One excites, and the other calms. One is "fight or flight," one is "rest and digest." One is the sympathetic, the other is the parasympathetic. Both connect to all our bodily organs, fighting for control. Throughout both networks are bundles of nerves called ganglions, which operate like mini brains.

With the extreme breath holding, Wim Hof is going hard after the sympathetic system. He's putting my body on high alert. First raising blood oxygen levels and lowering carbon dioxide levels through hyperventilation. Then reversing that by holding my breath. This sequence has a purpose.

I am my adrenal gland. Far above me in my brain, the nucleus of the solitary tract buried in the medulla oblongata is firing. An electrical signal arrives to me from Arvind's brain. It says, "Stressing out. Low oxygen! Careful!" I immediately release adrenaline to kick-start a fast defensive response. In milliseconds adrenaline spills out of my inner layer of cells into a dense network of blood vessels. The adrenaline hormones do their work almost instantaneously. They flood Arvind's body, acting on receptors in nerves, muscle cells, and immune cells. Arvind's heart rate spikes, his breathing rate spikes, and his white blood cells release anti-inflammation molecules called cytokines. White blood cell count will also increase in number over the next few hours.

"Feel the urge to breathe, it's OK!" Wim says in a calm but still booming voice. "BREATHE! FULLY IN!" he commands. I suck in deeply, and the sounds of the morning explode into my brain. Chirping birds, wind blowing through tree branches. Ants are crawling over my legs. I don't mind at all. I held my breath for over two minutes pretty easily. I feel euphoric but nothing crazy. Others around me are screaming with joy. Something very real just happened, just by breathing.

The ice water bath is next. I am worried. I have 4 percent body fat and hate cold. I keep the house at seventy-five degrees Fahrenheit. I freeze in a breeze. Wim is saying get into ice-cold water to my neck and stay there for two full minutes. It's a beautiful morning, midseventies. Ready. Set. I jump in and immediately freak out.

I am my parabrachial nuclei. I am deep within my brain stem. I am receiving a flood of signals that my temperature is way out of whack. I activate, firing a signal to my buddy the periaqueductal gray, or PAG for short. The cool thing about the PAG is it creates endocannabinoids instantly and starts pumping them into Arvind's blood. No, it doesn't get Arvind high, but it does help him deal with the pain. The immune system is activated again.

Breathe, breathe, breathe, I tell myself. In deeply through the nose, out the mouth. Focusing on a particularly shiny ice cube floating just in front of me, I relax, barely. The cold feels like millions of daggers piercing my skin over and over. It hurts. I try to stay strong. Feels like forever. Suddenly, Wim starts a countdown. Ten seconds. Five seconds. Four. Three. Two. One. Out. Thank God it's over.

There are grins and laughter all around me. Between the endorphins and the adrenaline, our friends are definitely feeling very alive right now. It's a bit like a nitrous oxide hit. I may be a little less affected by the novelty of it because of my intense MMA training and fighting, where it's fight or tap nightly. But Wim tells me that I won't get sick on my flight back home because I have fully activated my immune cells. I thank him, but not sure why.

The private Wim Hof is not really different from the public Wim Hof. He's definitely living in his primal brain. He's instinctive. Among a group of people who tried quite hard to be sensitive and not offend, Wim is like inviting a bear to the party. A bear who wants to hug you. He is unrestrained when he speaks, loud and intense. People give him slack because he's Wim Hof, he's all that we asked for. He can also go quiet for long times.

There's something fabulously retro in Wim Hof. On one hand, he's a New Age phenomenon, catnip for the overstressed masses. But he's also kind of a Marlboro Man, crossed with Gérard Depardieu. A bit larger than life. Sorta old-school. Get over it. Don't be a victim. Take control. Be powerful. Get outdoors. It starts with breathing.

A month later, back in my lab, I think about that beautiful Venice morning and try to unpack what happened to my mind and body. I didn't get sick in my travels, even though I was on the move for more than twenty-four hours through multiple airports. I am impressed. I think Wim Hof is on to something.

Civilization is, by design, a way to buffer the intense stress of living in the wild. So civilization's inventions largely make life more comfortable. Puffy jackets. Prebutchered meat. Umbrellas. Air-conditioning. Civilization today creates air-conditioned experiences, stripped of extremes. It has narrowed the breadth of human experience over millennia, shallowing the relationship between our bodies and brains.

Back when we had to hunt for meat and potentially be killed in doing so, our fight-or-flight system was engaged deeper and more often than today. Having a bad meeting at the office is just not the same as putting a spear through the heart of a wild boar.

The antidote to comfort is discomfort. That's why the Wim Hof Method needs the ice water. Only something *that* shocking can elicit a true fight-or-flight response. The breathing helps but the cold water is key.

Clinical studies on the Wim Hof Method carried out in 2013 support its effectiveness in boosting our immune system. By activating our autonomous nervous system through breath and cold water, we are able to reduce our inflammatory response and increase our white blood cell count. My friend's allergies probably cleared up by boosting his levels of adrenaline, a known antihistamine.

The Wim Hof Method has an ideological resemblance to the keto diet. Ketosis is how the brain protects itself during long periods of starvation. It's another system wired in for extreme situations; triggering it is, like Wim Hof, about reawakening that lost ability.

The most important thing to understand here is that Wim Hof's breathing method is the complete opposite of other breathing exercises you're probably more familiar with, such as the long, slow breathing of meditation and yoga. Wim Hof uses extreme breathing to hit the sympathetic system; traditional slow, deep breathing in pursuit of calm activates the parasympathetic system–its opposite.

Controlling the parasympathetic system with breath was the first and original breath hack. Yogis began trying thousands of years ago, trying to reach a kind of extreme calm. Tales of unbelievable physical endurance abound from that time. Tales of fakirs, a type of yogi, being buried for hours and rising back out of the earth forty days later like in *Night of the Living Dead*. Tales of yogis living on nothing but prana, energy from sunlight and air with no food or water. Is this possible?

Lukka, my four-year-old daughter, is watching Po struggle with the demons of his past. No, not Po Bronson. Po, the panda bear. From *Kung Fu Panda*.

Lukka and I are watching *Kung Fu Panda 2* when Po gets wrecked in a fight because, in a critical moment, he sees a symbol on the villain's clothes that reminds him of being abandoned by his parents. This stressful memory distracts him from being able to defeat the villain and live up to his potential, becoming the crux of the story. How can Po find inner peace? Master Shifu provides clues. "Suffering. Pain. Sitting in a cave with no food or drink for fifty years." Sounds familiar. In the end, Po finds inner peace by letting go of his past, embracing tai chi and meditating.

Everyone seems to be looking for inner peace these days. And Buddha pushers everywhere are telling you to give up worldly pleasure or chant in a room with thirty other lost souls. I don't want to feel like a mystic or feel like I'm in a cult to get inner peace. Mindful meditation and transcendental meditation are two popular approaches these days, even delivered in app format right to your phone.

On the movie screen, Po begins meditating, breathing slowly. His heart rate slows. His mind focuses inward.

Like thousands of yogis, secular breathers, and charlatans before him, Po begins a ritual. There are countless styles of breath, each with unique claims of their benefits to your mind. But there are only two ways it affects health. The sympathetic system is activated by fast breathing and breath holds. The parasympathetic by slow, deep breathing.

The physiological target in slow, deep breathing is the vagal nerve, a kind of octopus-like tangle inside us all. We're learning a lot more about the power of vagal nerve stimulation through Silicon Valley startups, which are using medical devices to tap the vagal nerve near the ear. By firing the vagus for a solid minute to four minutes, they're seeing miraculous effects. They are turning these brain gadgets into treatments for depression, addiction, epilepsy, and chronic inflammation.

I am Po's vagal nerve. I am the largest nerve in his body. I am so large I have four nuclei and bundles of nerves attaching to multiple organs. I connect his mind to his body quite literally. When Po says he thinks with his gut, he's talking about me.

I notice Po's breathing has become slower. Deeper. I notice it through nerve endings that tell me how stretched his blood vessels are. The more stretched, the bigger the blood vessels and the harder they are working. I don't see adrenaline, so I send acetylcholine to the brain. Let's relax. Chill out. With each breath I get more excited. When I get excited, Po relaxes more. Rest and digest, buddy.

Just when Po is relaxed, he keeps going. Po takes it slow to the extreme. He keeps breathing slowly and humming softly, which keeps me activated for more than twenty minutes. At this point I begin to message the insula. The insula is the heart of subjective emotional experience. It's the brain region where our perception of what's happening to us matters more than what's really happening to us. The insula releases a flood of serotonin, filling Po with a sense of well-being and warding off depression. I send signals to inhibit TNF-alpha, a cytokine, to stop it from being made. This has an anti-inflammatory effect in Po's body. Po is done breathing. I slowly stop stimulating his insula. But he keeps feeling great.

I walked onto that grass in Venice worried just how seriously we could take Wim Hof's approach to life. I'm left with this judgment: We should take it *very* seriously, perhaps more seriously even than Wim Hof himself understands. Breathing is key to accessing the unconscious neural code that controls us. "As the breath moves, so does the mind" is ancient Indian wisdom. We can add to that, "and the body."

More broadly, the real question here is evocative, and quite suggestive. How many more primal systems in our bodies have we ignored that can impact our health? To what extent can we truly heal ourselves? It's treacherous ground. People have died believing they can cure their own cancer without the help of sophisticated drugs designed on the computer with physics modeling. But there are many systems in the body that we are just learning to tap.

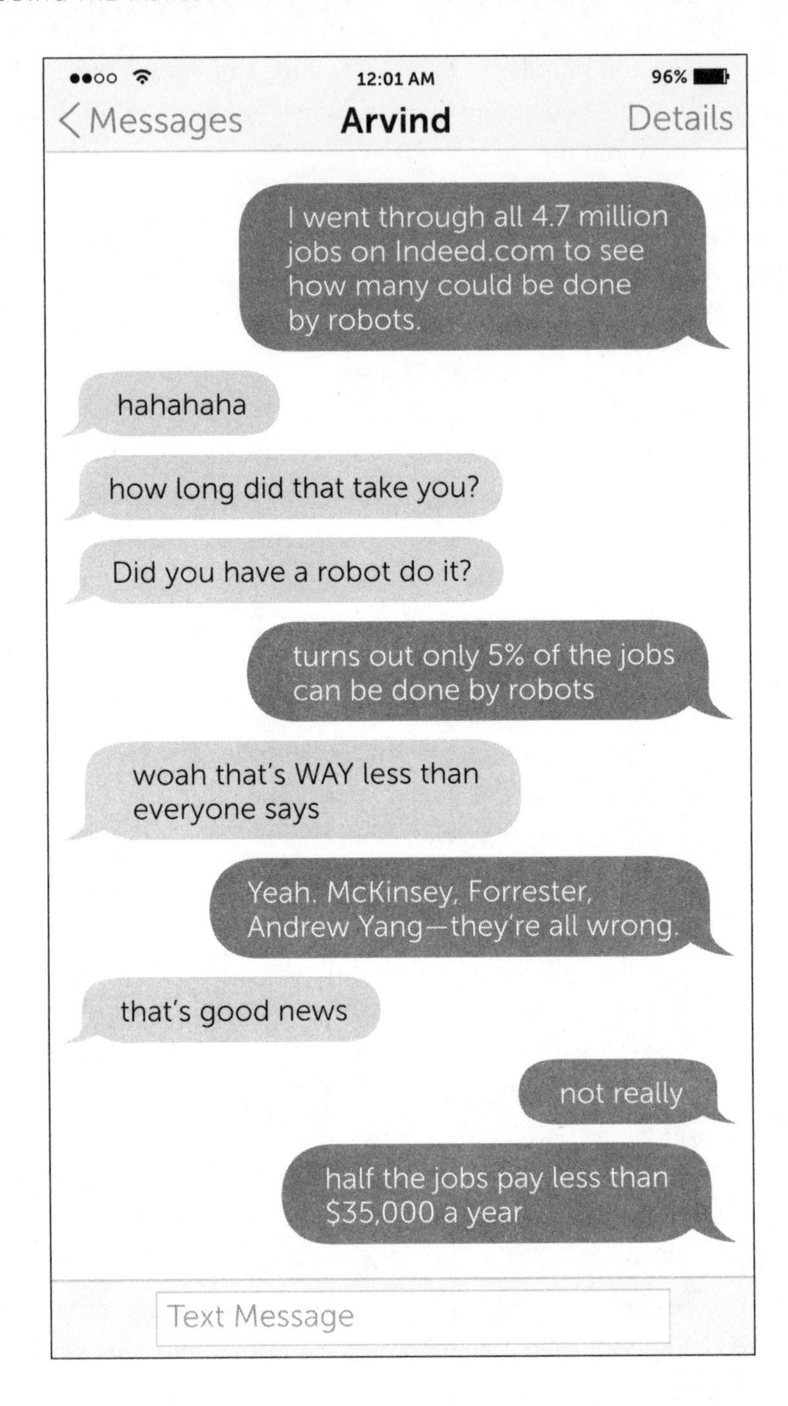
12:01 AM
96%
Messages
Arvind
Details
I went through all 4.7 million jobs on Indeed.com to see how many could be done by robots.
hahahaha
how long did that take you?
Did you have a robot do it?
turns out only 5% of the jobs can be done by robots
woah that's WAY less than everyone says
Yeah. McKinsey, Forrester, Andrew Yang—they're all wrong.
that's good news
not really
half the jobs pay less than $35,000 a year
Text Message

21

Walmart Will Soon Use Hundreds of AI Robot Janitors to Scrub the Floors of U.S. Stores CNBC

Growing up in Seattle, I started working in the summers sweeping and mopping floors. I worked at five restaurants and cafeterias, and then swept and mopped at a steel warehouse and a bus lift assembly line. In college I mopped the student cafeterias, the commissary, and the frat house. By the summer after my sophomore year, my long experience as a broom and mop expert helped me move up in the world. I joined the union as a floor man, working the night shift. My partner and foreman was twenty years older and very intense. At 10 p.m. we'd go clean out a few banks, then around midnight hit the Ash Grove Cement plant below the West Seattle Bridge. You might think a cement factory was more of a mess than a bank or a student union, but it was the other way around. It was a big facility, though. I had hours to think. If I ever slowed down, my partner would hiss at me. If I sped up and missed a spot, he'd hiss even more. This was his livelihood. It paid. He had disability if he needed it. It was consistent. If we got a complaint, it affected his reputation far more than mine.

So when I came to IndieBio, I took interest in some of the autonomous cleaning robots displayed upstairs at HAX, our sister accelerator, which has invested in–and helped make real–many robot companies. They've invented the robot shelf scanners for grocery stores, robot skyscraper window washers, robot wall painters, robot safety inspectors, and robotic trash/recycling sorters. They have a robot that goes in the hotel lobby and will wash and press your dress shirt for $5 in four minutes. But my favorite was the autonomous janitor *Neo*, an industrial mopper robot that cleaned floors for factories, malls, airport terminals, and the like. *Neo* is the same size as a human-driven, ride-on floor scrubber, a pint-sized Zamboni.

Seeing *Neo* on my way into work every morning brought back a lot of memories.

So Arvind and I decided to call the man behind our robots. His name is Duncan Turner. He and Arvind worked together in China before both joining our venture fund. The HAX office in China is above a massive electronics mall, where any spare part a robot designer could dream of is for sale downstairs.

We mentioned that a lot of people (us included) are worried the robots will take the human jobs away.

"Yeah, that's basically what I do." Duncan laughed dryly. He added that the companies we see today started out five years ago, or even earlier. It was very hard to get to this point, where they're having success. Most of the companies were building robots before society was ready for them.

Today, with new robots that will run 5G technology, it's easier. Robot designers can now count on the fact the robots will be permanently and completely online, and everything the robot senses can feed into the supercomputer in the cloud. Before this moment, decision-making had to happen in the robot brain. Which required a bigger computer on board. Now that we can make brainless robots, they're actually smarter, because more of the decision-making can happen in the cloud.

So next time you're watching Netflix on your phone using 5G, just keep in mind there's a robot also on the network, doing work our parents used to do.

The 5G connection also opens the door for a remote human to take over, from some command center on the other side of the planet, if the robot ever gets in a jam. Until now, robots had to do mundane stuff. They work best when people barely notice them. Creating robots that attract attention hasn't worked; they creep people out. Then there's robots that do "robot theater." Where people can watch the robot barista, for example. Duncan doesn't think these robots will last. Vending machines can make a perfectly good cup of coffee without a fancy robot arm flying around.

We asked Duncan what he looks for in a team of people who want to make robots to replace humans. His answer surprised us. "Not robot enthusiasts," he quipped. By that he meant kids who had spent their life in robot clubs at school. They tend to want to push the limits of what a robot can do, and they put in too many moving parts, and too much software. It makes the robot unreliable, or break down.

What makes a good industrial robot–one that can take the place of humans–is that it never breaks down and is extremely consistent. Otherwise, you're just replacing the human with something that's too humanlike. Humans break down. Humans aren't consistent.

He mentions one of his new companies has a robot who cleans bathrooms. Duncan calls it a fantastic example of using a robot to do a job that no human wants to do. There's a lot of dirty jobs humans don't like and don't do that well; quality and safety inspections is one of them.

Duncan explains that "humans are illogical. They create randomness." Humans don't pay attention very well. Robots like to operate away from humans. "Anywhere we can completely remove the humans, the robots do well," he says. Factories that run in the dark are an example. So is mining; he says the Rio Tinto corporation in Australia has more miles on its self-driving diggers and transporters than any autonomous car company.

The biggest challenge to robot companies, I soon learn, isn't the robots. It's the humans. To make a robot that works around humans requires a ton of sensors. The robot has to constantly monitor the human in case it does something unpredictable, which it does a lot. Robots have to be prepared to freeze at any second, or stop work and get out of the way for a moment. "If a robot is not programmed to work with humans, it can't work with humans."

We challenged Duncan that robots aren't needed when humans are so cheap. You can hire a human for only $9 an hour. Humans, you might say, are the first and oldest technology. Wealthy humans have been using poor humans to get work done since the dawn of time. "No, no, no," Duncan says. "Humans are much more expensive. They need to be fed. They get sick. They need sleep." They feel different from one day to the next.

What's different about the robots of today is that they're replacing service jobs, ones that couldn't be offshored. There are 3 million security guards, and 3.5 million cashiers. 8 million who do food prep. 4 million warehouse workers. 1.7 million truckers.

I just had to get one more question in. Does he ever get applicants from companies that want to make sex robots? He laughed. For sure. But his reason for disliking them was surprising. He described the sex toy business as an industry with a lot of people running their little fiefdoms. They don't like outsiders; it's inefficient. In other words, it's dominated by humans more than corporations, and because of that, it's infused with irrationality and unpredictability.

Wise-guy pundits find ways to talk about robots that make them less scary. They'll say, "Robots don't replace jobs as much as they just do a specific task." The *Neo* robot is one of their favorite examples; they point out how it can do floors but humans still had to clean the bathrooms. But even as they were making this fake insight, the robot bathroom cleaner was in development.

Another grandiose parsing of bullshit is the declaration "We don't want to replace humans, we want to make humans far more productive." It sure *sounds* nice. But replacing workers and making the remaining ones more productive are *literally the same thing*. If a team of three grapevine workers can now do the job of thirty with a $100,000 machine that does shoot thinning, pruning, leafing, and suckering, you've "made them far more productive." Machines (and tools) have been making humans more productive for thousands of years.

The biggest lie of all is the notion that robots don't take jobs, they create jobs. The robots don't do that–the humans do.

We like Duncan because he doesn't fall into this pandering and wordsmithing. He's honest and doesn't mince words. He doesn't manipulate perception.

Some people are confident there will be new jobs in the future, but we can't imagine them before they've been invented. "Look at how many yoga instructors there are," said my colleague Kevin Kelly. "Twenty years ago, nobody would have imagined we would have so many yoga instructors. Back then, who would have thought you could make a living as a yoga instructor?" I understand the point, but I also think it's not very satisfying. (There are 52,000 licensed yoga instructors in the United States, about half of them active. Meanwhile, 6 million people work as janitors and in grounds crews.)

I don't buy that the future is unknowable. I tend to believe the future is always already here, it's just in code.

So, I went in search of the jobs of the future. Not the jobs that people with graduate degrees in engineering will have. Rather, I went looking for what will happen to the millions of people displaced by robots. I arrived at a rubric to understand what that life will be like. It's not a jobless future (unless there's a recession). This surprised me. But it is distasteful; it's merciless. It coughs up some painful truths that crash headfirst into our ideals.

1. THE NEW TECH INDUSTRIES AREN'T THE JOBS SOLUTION

This is sort of self-apparent but, I think, overlooked. We actually do know what many of the industries of the future are. They're being invented right now. They've *already* been imagined. And the people they will hire are mostly PhDs and engineers. Not many laborers. Ninety-nine percent of the "new industry" jobs created since 2010 have gone to people with a graduate degree.

Virtual reality is a future industry; it doesn't need many physical laborers. Nuclear fusion is a future industry; it will need fewer jobs than the coal and gas industry it replaces. Cryptocurrency and blockchain don't create jobs. These technologies just make it easier to get paid in fractions of pennies. Even biotech is rapidly adopting robots to replace lab techs. All of the future industries will follow the Power Law, which is VC speak for "winners take all."

You can't point to a single deep-tech field and argue, "That's going to create a lot of jobs for everyone." Sometimes politicians point at solar power and say it will create jobs; certainly the first-time installation takes some labor, but not after that. And these jobs pay less than the job they replaced (which is a big reason why they're more cost-efficient). A coal miner in West Virginia averages $64,000 a year in salary. A solar-installer job pays about two-thirds of that, around $40,000 a year.

What these new industries are good at is turning those graduate degrees into cash. They crank out lots of prosperous people. Tons of them have incomes over $100,000, which is the eighty-fifth percentile of income in the United States among all workers (which doesn't even count the huge number of retirees, disabled, and students). If you have a house with two adults, and in any combination they earn $180,000 a year, that's the ninetieth percentile bracket for households.

And it turns out this high number of prosperous people is critical to keeping everyone else employed.

2. THE "STRESS ECONOMY" SPREADS THE MONEY

If you follow the money, you really start to grasp what creates jobs for the thrift class. But it doesn't start with the thrift class (they don't have the money). It starts with the prosperous. Imagine twenty urban households, each with a couple adults who have graduate degrees and earn very good salaries. Every year, their houses go up in value and their savings, invested in the stock market, also go up in value. These are the people earning the lion's share of the money.

But their lives are very busy. Their children are overscheduled. Stress is their number one problem. And they have the discretionary income to try to solve this problem. This simple dynamic creates a massive demand for service workers in the thrift class.

Rather than compromising their careers, they hire a nanny. Rather than taking the bus to work, they might take an Uber. Rather than endure the grind of cooking dinner, they go out to eat (or order takeout). Rather than stay home on Friday night, they hire a babysitter. Rather than mow their own lawn, they hire a landscaper. Rather than coach their kid's soccer team, they pay an expensive academy to handle it. For a treat, they get a pedicure.

And the more these families earn, the more they can spend on reducing their stress. One of their big stressors is keeping up with the Joneses. It stresses them out to think of their children trying to learn in a public school, where there's thirty-five students to every teacher–so they send them to a private school, where there's only fifteen students per teacher. They hire admissions counselors and tutors and test-prep instructors. Just organizing all this causes them even more stress, so they pay massage therapists, and yoga teachers, and psychiatrists to get them back in the groove.

So in a way, Kevin Kelly was inadvertently on point when he said the jobs of the future would be the equivalent of yoga teachers. The jobs of the future are all taking care of the well-off. There's almost no end to it. One of the biggest stresses in these households is when things go wrong. The house has ants, or the front door lock is catchy, or they get in a fender bender. When they call customer service, they want to talk to an employee of the company. Not an answering service. Not someone in India at a call center reading a help menu.

This may sound like a very cynical way to view the economy, but it's accurate. You can clearly see this on full display in the job listings on the online job boards. When I started writing this chapter, Indeed.com listed 4.7 million jobs. Seventy percent of them paid less than $50,000, and half pay $35,000 or less. Only one out of seven of them paid $70,000 or more. Nine out of ten jobs were in cities. Virtually all the jobs (97%) were in the service economy, taking care of people and their things (and their bodies) and fixing them when they break. Only 3% of the jobs were in manufacturing. (Surprisingly, I saw two of our biotech companies listing manufacturing jobs; one paid $20 an hour, the other only $15 an hour.)

The good news was, not many of these 4.7 million jobs looked like robots could do them. The cashiers can be replaced (like in an Amazon Go store), but only 1% of the jobs were cashiers. Truck drivers were also about

1%. Some of the inspection jobs could be "made more productive" with robot help, but those jobs were below 1% of the whole.

For the most part, more than 95% of the jobs listed required the one thing that Duncan described robots as being bad at: working around humans in a complex way. A robot can't come to your house, test for insects or rodents, and spray pesticide. A robot can't deliver and assemble furniture in your house. A robot can't supervise a doggy day care center.

3. THIS "STRESS ECONOMY" IS A HOUSE OF CARDS IN A DOWNTURN

The vulnerability in having an economy like this is that so much of it is discretionary. It doesn't *have* to be paid every month, like a mortgage or a utility bill or a car lease. A dangerously high percentage of the spending is highly dependent on the prosperous feeling flush. As long as they feel wealthy and stressed, the money keeps flowing. But it would be really easy to cut back if needed. Eat at home more; have the landscaper show up less often; go without a massage.

When the pandemic hit, the number of jobs listed online dropped by 2.1 million. It was an unprecedented shock, and it called attention to how many jobs are actually fungible.

So the conclusion we can draw is that the biggest risk to employment isn't robots. It's recessions.

So can robots cause a recession? Absolutely–if they displace too many workers too rapidly. Out of work, people would freeze their spending, and this would work like a price shock, sucking the economy of the usual monthly bolus of cash. But gradual robot adoption aids productivity, lowering prices–robots are like lending: good in moderation. And I still would say that bankers are far more dangerous than robots.

4. HOW THE OTHER HALF LIVES

There used to be a phrase. "How the other half lives." Well, it's way more than half.

Seventy percent of society now earns less than $50,000 a year, and has less than $500 in savings. This "thrift class" is projected to increase, not decrease. The middle class is disappearing.

So if you have this idea that in the future, a lot of people will have to survive on very little income–well, that'd already been happening. Long before the Great Cessation.

Silicon Valley is both the push *and* the pull. Automation, robots, machines, offshore labor, international plane flights, video conferencing–all of that is the push out the door. The pull is the other side of the equation. At the same time as we made it harder and harder for regular people to earn money, we also used our mastery of technology to create things that made it easier and more enjoyable to live on less money.

Thrift means, simply, "being careful with money." Living that way is not new. What's different today is how many conveniences of modern society have, critically, made it easier to stretch a dollar–which, in turn, has allowed the thrift class to grow.

We now have tons of retailers who compete at IKEA prices. Sharing economy platforms defray the cost of having to buy things. Thrift-store shopping is now a $7 billion business. Cheap, unlimited entertainment like Netflix, Spotify, and Fortnite is filling thousands of hours of people's time. People read so much now, for free on their phones, that we call it an addiction. As one thrifter texted me, "It's never been so entertaining to be poor."

The thrift class has shown incredible imagination and creativity in adapting to the circumstance and making it their own. What's fascinating is how they've turned the tables on their predicament, building a moral philosophy around their lifestyle. There's a rejection of consumerism. It's liberating to get out of the trap of always earning more money and buying more things. Spending a lot of money on something–when there are savvier alternatives–is increasingly seen as gauche or just dumb. In a world where the jobs aren't offering enough people a sense of self-determinism through work, they're meeting their need for self-determinism and freedom through a kind of minimalism.

And while the power players and policy makers are trying to figure out how to save the planet from climate change, the thrift class has its own solution that's already fully up and running. It's called buying less and sharing more. Rather than pollute the skies with CO_2, just don't buy a car at all–and stop going everywhere this way and that. Rather than flush microplastics into the ocean with every wash cycle, don't buy new apparel at all. Rather than jam more plastic bottles into landfills, just drink tap water. Buy beer in refillable growlers. Don't *buy* clothes at department stores–just take the clothes into the dressing room, post a photo of yourself on Instagram wearing them, and leave as the likes start accumulating by the hundreds.

In the thrift-class way of viewing the world, freedom isn't something you reach by working for decades to earn a big pile of money. Freedom comes from being master of your own time. Gig work is a trade-off most are willing to take, because they can prioritize their personal lives. If they want to hang with friends on Tuesday, they just don't take a gig. They've seen their parents run ragged from trying to "have it all," and they don't want the stress.

On Abraham Maslow's pyramid, nobody knows where they are anymore. The thrift class is taking care of lots of its wants before taking care of its needs. It works only until it doesn't. All it takes is one medical bill to trigger a downward spiral.

Millennials are spending $20 less per day than their counterparts ten years ago. College enrollment is down 5 percent this decade. The birth rate is down 15 percent. This is perhaps one of the most wincing details of our times. People with tons of time available to parent kids but too little money to feel safe having them.

5. CHANGING IT IS IMPOSSIBLE, BUT IMPROVING IT IS EASY

Income inequality in the United States is obscene. There are a lot of suggestions to remedy it. They all take money from the prosperous and give it to the less fortunate. But among these options, are some better than others? Absolutely.

Huge medical bills destroy lives.

(Medicine can solve this, and we address it in Chapter 28.)

Make state colleges free.

During this era where we knew that command of new technologies would be important for the future workforce, we engineered the system to *reduce* college enrollment. We raised the price of college so much that students are burdened with record debt. In this economy, a bachelor's degree is not enough to climb out of the underemployment trap. It doesn't cut it. The number one way to give people a chance to be prosperous is getting advanced degrees. But we've rigged the system such that only the children of the prosperous have that chance.

Add a tiny tax on wealth, rather than raising the tax on income.

We built a system of amazing wealth creation, called "venture capital," to profit on future technologies like robotics and biotech. But we forgot to make sure that everyone, no matter how poor or rich they were, had equal access to it. Anyone can become an entrepreneur, but you need a million dollars in the bank to qualify as an "accredited investor." So these amazing gains are only for people who are *already* millionaires.

As their shares accumulate in value by the billions, *investors are not taxed.* They're only taxed if they sell their shares. But there is no reason to sell the shares, because they can easily borrow cash and use the unsold (and untaxed) shares as collateral. In this way, billionaires are literally paying zero taxes. They can hold the shares for ten years–or fifty–and never pay tax. And if they do have a small tax bill, they can donate some of those shares to their children's university to offset it.

Raising the minimum wage.

This is a one-hundred-times better idea than Universal Basic Income. You can put $10,000 in a worker's pocket just by raising their wages $5 an hour.

What we can't do is change the fundamental nature of our economy. If the new technology is "winners take all," then the economy will increasingly shift in that direction, too. We rely on the prosperous. It's ugly, but it's real. Everyone else is discretionary.

●●●○
11:25 PM
45%
Messages
Arvind
Details
Did I ever tell you about Crypto Joe? Our first security guy?
here at IndieBio?
yeah he got into Bitcoin—while he was here
how much money did he lose?
No he's worth millions. He's been in the New York Times.
You're telling me that our old doorman...is now a Bitcoin whale?
Yeah. Hilarious!
Text Message

22

Billionaire Warren Buffett Calls Bitcoin "Rat Poison Squared" Coindesk

Disclosure: Our venture fund SOSV has a 6 percent share of a crypto-derivative platform, BitMex. BitMex is not authorized for use in the United States, because it's not registered with our financial authorities. If you think buying Bitcoin is risky, BitMex allows you to bet on whether the price of Bitcoin is going to go up or down–and multiply your bet by up to one hundred times. So with a mere $100, you can make a $10,000 bet. And if Bitcoin goes up just 10 percent, you can make ten times your original money. (Technically, the synthetic law-splitting mechanisms of crypto make this not a "bet" in the legal sense.) Of course, BitMex is wildly popular among one of the two types of crypto investors–the type that wants to get rich quick.

But there is another type of crypto investor. They got into it for the long haul. They now hold vast amounts of wealth in crypto assets. They tend to be deeply philosophical and have very elaborate mental models of nation-states, global monetary policies, and the role of finance in keeping political economies afloat. They don't buy and sell cryptocurrencies for short-term gain; instead, they watch the daily transactions the way a seismologist studies vibrations of the earth for signatures of future earthquakes. Their minds are preoccupied with signs of war, government instability, price shocks, and abuses of power. They use the phrase "asymmetric bet" frequently. They don't have faith in Bitcoin so much as they feel it's a smart, asymmetric bet. They admit quite candidly that there is a fairly good chance they will lose their money–but there is a small chance that Bitcoin will go up fifty times its value today.

These are people who easily could have cashed out–coming back to the real world with their tens and hundreds of millions in profit and living the good life. But they made the decision to stay in their synthetic/imaginary world, like Peter Pan, and keep fighting for their long-term vision of a superior global financial system.

We found this pretty interesting.

And so for this chapter, we sat down with a bunch of them to hear what's on their minds, to try to understand why crypto isn't just another dot-com bust in the making. Some didn't want to be identified (and why that would be the case we'll explain). Some are old acquaintances. Most of them now run crypto funds on behalf of big investors. Most of them also have a background in computer programming, though not all. One of those was Crypto Joe. He's an "only in Silicon Valley" fairy tale. Arvind will tell you about him.

The first time I met Joe, he hit me flush in the face with a left hook. At the time he was known as "Tall Joe" and we were at the gym, MMA training. Joe was a poor kid from the San Francisco peninsula. I'd just moved back from China, and I was building the IndieBio office. Given our rough location on Jessie Street, I needed a doorman who could be muscle as well. Joe was perfect. He stopped driving for Uber.

Like a lot of kids, Joe felt like it was a joke trying to save money and earn pittance interest until his retirement. He was a hustler always scheming a way out of being poor. He'd worked for the Rand Paul for President campaign in 2012, and that exposed him to a lot of libertarian thinking about the financial system. He had been on 4chan, the dark web, looking for vintage porn to buy and sell for profit. He went on Silk Road and saw people using Bitcoin, and he thought it was more of a PayPal-type thing. But as he learned more, he started buying it, a little every month.

One night, here at IndieBio, Tall Joe hosted the first-ever Ethereum Meetup Group for San Francisco. Thirteen people showed up, but one of them was the CTO for the Ethereum Foundation. Ethereum was almost out of money, and at the time nobody knew if Ether would survive. But Joe bought a bunch because he thought smart contracts made sense. Joe took a coding class in San Jose and started buying into ICOs he thought would have a real market. He started to become an insider, and that was when he stopped being the doorman at IndieBio and became Crypto Joe.

Now he's a whale. A small whale. By 2018, he was in the *New York Times*, which skewered Joe for renting an orange Lamborghini for the weekend and wearing a solid gold Bitcoin "B" necklace ringed with diamonds. Joe told the *Times* his holdings were in the "double digit millions." Joe has a friend who sank half a million into Ether in the early days, and now has around $700 million. Joe is really chill and actually really serious about where crypto is going, and is a vocal advocate–he's met with the SEC and other regulators to explain why the new crypto systems being used today address some of their concerns. He understands the problems coming from proof-of-work and is promoting proof-of-stake as an alternative. He gives us advice to pass along to BitMex. Joe is twenty-nine now, and I still see in him that passion he always had. I asked him to spar again when we last met, but he's too busy making money.

So we've done these interviews with the whales, including with Crypto Joe, and I'm thinking we should rip off what we think are our most interesting impressions and observations.

For me, the biggest takeaway is that–on some deep fundamental level–crypto is owned as an asset hedge against the vulnerabilities of our current global financial system.

It's a way to park cash in case governments overprint money to rescue their economies and just trigger a meltdown. That's the core critique. When the Great Cessation happened, at first Bitcoin fell even faster than the stock market. It really looked like the believers had panicked and lost their nerve. But a few weeks later, when so many world governments created stimulus packages and started printing money by the trillions...the crypto scene went ballistic with moral outrage and disgust. And along with that outrage, the price of Bitcoin rose again. The more the governments bailed out the economy, the weaker those governments looked (financially). In the course of a few weeks, Bitcoin reemerged as the only way to bet against the fiscal policy of all major governments at once.

It's almost a dystopian play; it's rooted in a core belief that power *always* corrupts, and central governments–with absolute power over their monetary system–will abuse (and have abused) that power. Havoc and chaos are crypto's friend.

What happened in Cyprus in 2011–when their financial system failed, and everyone tried to buy Bitcoin to get their money out–is the prototype for what could happen to much bigger countries.

But within that pessimistic mindset, it's actually zealously idealist, offering not just a short-term solution to where to put your money, but a long-term solution for how to organize and run the financial system. Instead of centralized systems, which are vulnerable to power abuses, it offers a decentralized system, which nobody can singularly control. It's eerily similar to the internet itself. For better and worse.

In fact, the whales aren't that worried if Bitcoin goes to zero. That surprised me. All the Bitcoin in the world is worth about $134 billion right now. That's about the same as Philip Morris is worth. Just one big company. And if Philip Morris went bankrupt tomorrow, its shareholders would take a loss, but they wouldn't go broke.

For me one of the biggest revelations is how many top VC firms have invested a decent chunk of their fund into crypto coin. Not into crypto startups. I knew several VCs with a third of their *personal* wealth in crypto, and I knew some lower-tier VC funds who tried to boost their firm's results with Bitcoin. But I didn't really grasp until now that many of the top funds in the Valley are holding big positions.

You might think this isn't what they're supposed to do–that they're supposed to invest that money in startups–but at a certain point it became logical to do it. More than that. As one of my friends told me, "It became irresponsible for us *not* to do it." And the reason is, they're holding so much money, often billions, and almost all of it is in U.S. dollars. "The dollar is not riskless," he persuaded his firm. "That's a silly idea." And by having a moderate percentage of their fund in crypto, it's a hedge.

He told me, "Either the dollar will be stable, in which case our regular business of investing in startups pays off. Or the dollar will be in trouble, in which case the crypto will pay off. Between the two of them, one is going to work."

Bitcoin is the ultimate asymmetric bet in the Valley, and venture capital's job is to find and invest in asymmetric bets. "It's VC catnip," quipped a whale friend.

Crypto Joe talked about being at conferences where university endowments have announced they're now holding crypto for the same reason–simply that it's a smart hedge, an uncorrelated asset to everything else. It's downside protection. They own it for the same reason they put money in gold. Crypto wants to be digital gold.

It's quite revealing that this many serious investors, who manage other people's money, institution money, feel the need for a hedge against the American economy going into a freefall.

The fact that so many big players are in the game is important to its future legality. The whales were honest with us–the likeliest scenario is that the U.S. government declares crypto illegal, because it can't do enough to satisfy the regulations against money laundering in the post-9/11 era. Nobody wants to make it easy for terrorists and crime networks to operate.

So the hope is that it gets too widespread for the government to shut it down. Sort of a "too big to fail" argument. The more that wealthy, well-connected institutions hold crypto, the less likely the U.S. Treasury Department is to do something harsh that wipes out those holdings' value.

One of the whales described the key problem the Feds have. "If North Korea hits my node for a transaction, I have no idea it's them. Even if I knew it was them, I couldn't stop my computer from servicing their request. The whole system is permissionless, which also means it can't keep anybody out." In the Bitcoin universe, nobody needs permission to act like a bank or a broker-dealer. They just need the software.

However, Bitcoin is completely out in the open. It makes a public record of every transaction.

And because of that, we were told, criminal networks have left Bitcoin for alternative coins like Monero, which is an obfuscated ledger, meaning no user can be tracked or traced.

That Bitcoin is no longer favored by criminals has made it more respectable, and led far more investors to get in.

But nobody talks about their Bitcoin wealth out in the open. "You can get killed for that," one whale said. He meant that whales can be targeted by people wanting to steal their password.

There are wallet protection and insurance systems. But because they are hackable (in a way that Bitcoin itself is not), no whales use them.

You can't understand crypto without first understanding why it was created. There were weaknesses in our financial system, most of which are still there.

One of our whales had been a cryptographer in college when crypto was postulated. He read the original Satoshi Nakamura white paper and thought it was a really elegant system. He loved it for its elegance. Later, when Bitcoin turned on, at first it was free. Someone emailed him a link to the "faucet." Bitcoin had to spread around to start to be useful. He gave half his away to a friend, who then lost his key.

The reason Bitcoin is seen as unhackable is there's no central server to hack. Every node on the system—and there are about ten thousand—keeps a full copy of the software, sort of like each cell keeps a full copy of the whole genome. A hacker would have to break through security protections in over half of the nodes simultaneously to take control of the system and change its code.

But Bitcoin has a very serious crisis of conscience coming. The core problem with Bitcoin is that it's become a silicon arms race. The way it's designed, it rewards faster, more powerful computers. So everyone has raced out and bought expensive GPUs and tried to use their university's supercomputer to exploit it. It's sucking tons of electricity. That's made it way less democratic, and less decentralized. Very typical human things have happened, such as these computer miners forming pools. The computing power is getting concentrated into these oligarchy-like unions. If anyone hits 51 percent of the total computing power online, they can take control and change the rules.

Because this situation betrays human ideals of decentralization, it's led to a newly proposed solution–the one that Crypto Joe is a spokesperson for, called proof-of-stake. It's like when your parents cosign your first mortgage; someone with a lot of coin vouches for new players with less coin.

Let me revise that; they're not agreeing to pay your debt long-term, it's more like your parents would be your escrow company, putting up their own money *only for the length of time it takes for the transaction to go through.* This method would end the arms race, but making this paradigm shift would be too radical for Bitcoin to go first. So other crypto assets, like Ethereum, will do it first.

The Bitcoin community is pretty convinced they need to find a new solution. But the community is very divided over what the new solution should be.

So let's summarize a little bit.

1. Bitcoin has no intrinsic value.
2. But big money needed an alternative to the dollar.
3. So they got in, drove the value up, and tried to consolidate the market to control it, like they always do.
4. This betrayed the ideals behind crypto in the first place.
5. Crypto now has to choose its future and can't agree what it should be.

Yeah, that's about right.

It's important to understand that even though Bitcoin holds no intrinsic value, it has a feature that no government currency does. Not even gold does. Which is its finite scarcity. The limit is set at 21 million coins. It will never pass that. Investors love that. That's why they hold Bitcoin and not these alt-coin knockoffs.

It's in stark contrast to dollars. Every year, the government prints more dollars, diluting every dollar out there by that much.

In whale culture, the ultimate honor is holding on for the time Bitcoin's scarcity drives its value through the roof. They call the ability to hold on forever "diamond hands."

Explain to me why it's not just a Ponzi scheme–why its value isn't solely driven by its hype and scarcity.

So let's contemplate a version of the future where Bitcoin replaces gold as extranational value—where enough investors agree that because of its true scarcity, it's better at playing the role of gold than gold itself. Gold is a pretty nifty metal, but its monetary value far exceeds its inherent merits. It's expensive mostly because it's rare, but even then it's continuously mined and the supply goes up. Gold does an *okay* job being "gold," but it's not digital—it's cumbersome.

All the gold in the world is worth $7 trillion right now. If Bitcoin replaces gold, that would put Bitcoin's price at a minimum of $333,000. Which is thirty times its price today. So the smart money in Bitcoin today doesn't really care if the price fluctuates between $10,000 and $20,000. That's peanuts to them. They're holding it for the big score.

And that's the *low* end of the scenario. Because once Bitcoin replaced gold, then it would be an extranational currency, the go-to investment that shoots up in price when nation-based economies go down. As the first truly global currency, each Bitcoin could be valued in the millions.

Right now, when any other country's economy looks weak, money *flocks* to the U.S. dollar. The U.S. dollar is considered the safest economy to store value in. But what if you think the U.S. monetary policy sucks? What if you think the U.S. is digging itself into a hole? You need somewhere to go.

That's right. That's what smart money wants Bitcoin to be.

It doesn't actually make sense to me—I mean it does in pieces, but when I put those pieces together it doesn't add up. The less people trust the world's going to be okay, the better Bitcoin does. But for it to keep its value, you have to trust that other people will still want your Bitcoin tomorrow. So it doesn't get you around the societal trust problem.

Then there's the intellectual moralistic stance against financial irresponsibility. Every whale thinks the U.S. government is overspending. They say Japan is the worst overspender of all, and now China is overspending, too. But they don't reconcile that with the blatant reality that crypto has become a den of fiscal irresponsibility with all the day traders. Guys are flashing cash like you wouldn't believe. They're using their stake as collateral and borrowing dollars and going crazy—spending it, or making futures bets and margin bets. And the whales aren't immune to this. Plenty of whales play "whale games," pumping up and knocking down the coin prices.

Whales and bitcoiners aren't married to Bitcoin. They are married to a world that is not hackable. A blockchain that makes the financial fair and transparent tools.

That's the part that has me excited. I don't know who said it, but there is a great quote: "When goods stop crossing borders, armies will." It's so true. Beyond wars, protectionism creates poverty. Grain rots in barrels. Unemployed workers in Mexico die in the Texan desert trying to get work in America. Resources like food, labor, and capital cannot get to where they are needed. A global frictionless currency is required to have a world without borders and ultimately a peaceful planet.

I don't know if it will be Bitcoin to unlock this vision. In fact I doubt it. But globalization cannot be put back in the bottle. And a system like a blockchain will help power a global way of thinking and living.

You know, some people think the world is going in the wrong direction, so they vow to change who is running the world. But other people–the kind of people in love with crypto–think that person is just a product of the system, and so they devote themselves to creating a better system. Bitcoin was pretty audacious because it's tried to reinvent money itself. The system runs on money, so if we reinvent money we change the whole system.

Mankind's greatest illusion was money. Money doesn't actually exist, it has no inherent value. But in an organized society where we all buy into the illusion, money is realer than fists, realer than bombs.

But there's something about crypto I find strange, even reptilian–and the only way I can describe this is crypto's approach to trust. Rather than improving trust, crypto has tried to minimize the need for trust. To create a system that requires no trust. To build a world that will still function even when all trust is eroded.

But we are mammals. Trust is how we're wired. In biology, trust between mammals is defined by hormones, by the presence of oxytocin and vasopressin, and their impact on the genetics of the brain. Emotions are neural algorithms. Sixth senses. Bitcoin is trying to peddle a reptilian system to mammals. You can tell me a hundred times, "Trust it," but if my brain is suspicious–you can't argue with that. You can't rationalize it. Trust is biological, not rational.

So it's not that we can't have the programmers build a trustless world that works perfectly. It's that we don't want them to.

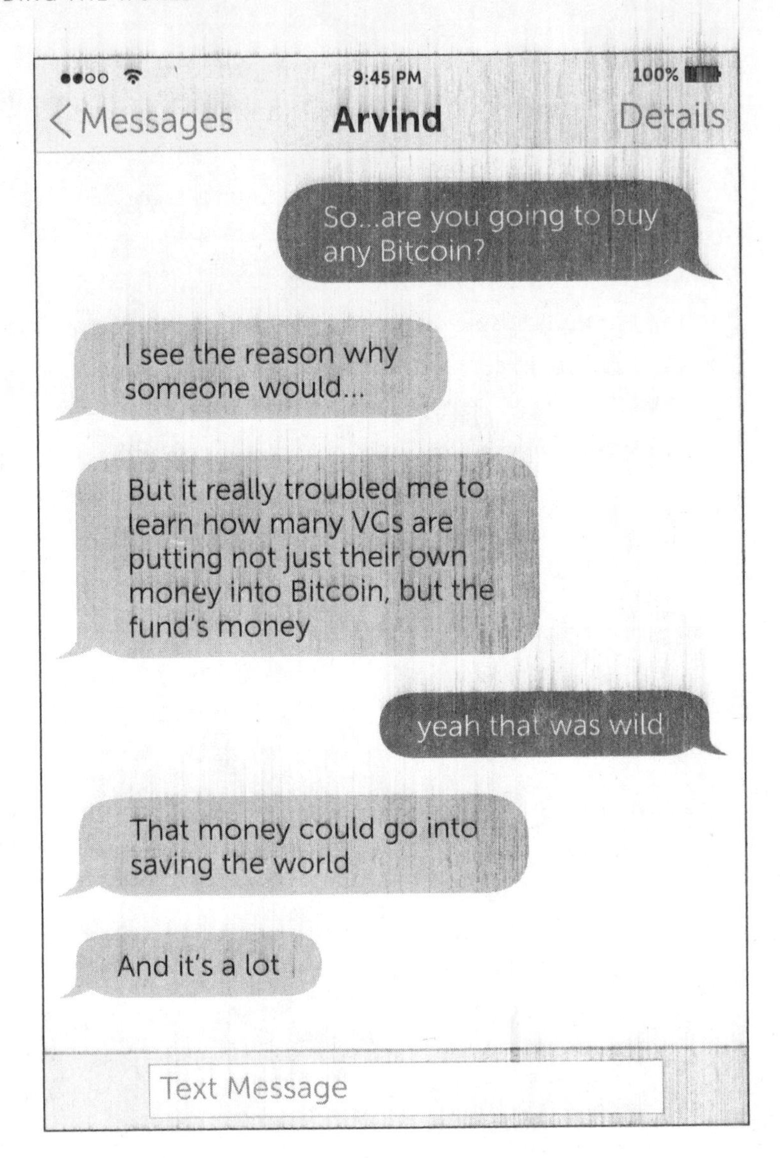
9:45 PM
100%
Messages
Arvind
Details
So...are you going to buy any Bitcoin?
I see the reason why someone would...
But it really troubled me to learn how many VCs are putting not just their own money into Bitcoin, but the fund's money
yeah that was wild
That money could go into saving the world
And it's a lot
Text Message

23

These Countries Are All Building Brand-New Cities World Economic Forum

One night in Paris, in 2004, I was walking through St. Germain with my family, when we were struck by a piece of art in the window of a gallery. The artist had recreated the civic architecture of the "ghost city" of Pudong, outside Shanghai, in pure white cardboard–then taken a huge photo of it, nearly ten feet by ten. Pudong was oft-ridiculed for being built in the middle of nowhere by China on the premise of "If we build it, they will come." But they had not come, and the city was empty.

We pressed our noses closer to the gallery glass, when we realized the art depicted a murder scene. On the roof of a building–so small at first you didn't notice it–there was a body in a black suit, dead in a pool of blood, with a woman in a black dress standing over him.

We really liked the work, and we took photos through the glass.

We got our moment of entertainment at China's expense, then we forgot about it.

Today, Pudong has 5 million citizens. It's no ghost. No joke.

China went on to build one hundred other cities, from scratch. They found it was a lot easier to build a new city on open land than to build onto an existing town. Not all of them are as successful as Pudong. But in the process, China got really good at making something the rest of the world now desperately needs: entire cities.

Roads, bridges, transit systems, energy plants, housing, office towers, dams, railways. Plus all the sensors and switches and cameras that go in them.

In America, we keep worrying about how Earth is going to have 9 billion people by 2050, or 11 billion, or whatever number best strikes fear. It's the wrong statistic to pay attention to.

What really matters–and what will truly transform the world–is urbanization. Around the world, every week, 3 million people move from rural areas into cities. That's a San Francisco every two days! That isn't a *future* statistic. That's a *today* statistic.

Existing cities can't handle the massive influx of people, so governments all around the world are mapping out land areas where an entirely new city could be built.

But they don't have the money to build a city in the middle of nowhere. Nor do they have the expertise. In many countries, they know if they tried to build the city themselves, corruption would be too insidious for the city to ever get built on time and on budget.

So who do they call?

The city-building expert.

While American companies were perfecting the operating system for our phones, Chinese companies (and their government) were perfecting the operating system for entire cities.

China is playing the game in another dimension. They are kicking our ass right now in global economics.

FOREIGN COUNTRY (LIKE UGANDA, OR MAYBE CHILE): We don't have the money to build a city.

CHINA: We will loan it to you at low interest. It will pay for itself.

FOREIGN COUNTRY: How will it do that?

CHINA: You will hire us to mine your chromites, your copper, your potash, your sulfur, your iron. We will bring it to China for processing and sell it for you. The loans will be paid off in no time.

FOREIGN COUNTRY: Wow. Maybe we should get two cities.

CHINA: A second city is a 5 percent discount.

U.S.: When it's ready, we would like to bring in a Cadillac dealership! And a Tiffany!

FOREIGN COUNTRY (IGNORING U.S.): We don't have trained workers to build a whole city.

CHINA: We will fly them in.

FOREIGN COUNTRY: Our people care a lot about the environment.

CHINA: A green city is 10 percent extra. You like solar or nuclear? Just check the box when you order.

FOREIGN COUNTRY: I hear you have a social credit score system for all citizens?

CHINA: Yes, that comes built in to Alibaba's "City Brain" operating system. Your city will have cameras everywhere. You will know what all citizens are doing, all the time.

FOREIGN COUNTRY: What if we want to block Google and Facebook?

CHINA: That'll be an option on your control panel. Our bots are very good.

FOREIGN COUNTRY: All right, I think we're all on the same page. United States, you bring in that Nike store, that Levi's store, the Tiffany, and fifty Cadillacs. China, you build our roads, bridges, ports, dams, subway, bus system, our entire downtown, and all the housing.

Most people have no idea this is happening at such large scale. China is investing $1.3 trillion in the next seven years in other countries—loaning it to them, outright, for the massive business of building entire cities. Which are run by Chinese companies, putting Chinese workers on the job. They have 125 countries signed up for these loans, stretched out to the year 2050. With their loan program, China has basically created its own alternative to the World Bank.

So, think about it. There are 195 countries in the world. China is loaning tens of billions of dollars to two-thirds of them for all the hardware that cities run on. Along with it comes China's software, its "city operating system," that keeps people in line. Also part of the package: nobody criticizing China. Or else.

Americans live in the only country that goes to such great lengths to protect free speech. In every other country, some aspects of speech are subject to liability. Or their constitution gives the government the authority to pass some laws restricting types of speech. So while Americans are appalled by the lack of civil liberties in China, a lot of other countries (I'd guess 125 of them) think it's pretty normal.

Everything in China is heavily influenced by Confucian philosophy, including speech. If guided by Confucian principles, citizens shouldn't be mouthing off on Twitter about national matters in the first place. They should be taking care of their own business, and their family affairs. Which includes (actually, prioritizes) looking after your parents. Maybe participate a little in local community affairs–but only after you've taken care of everything at home. So this is why most Chinese aren't that troubled by the idea they shouldn't be vocally criticizing national policies on the internet. That's been their tradition for 2,500 years.

Confucius would be appalled by America. To the legendary philosopher, we would appear to be in chaos. So much disagreement. So little order.

In March 2000, when Bill Clinton put forth the China trade bill, he declared, "China is not simply agreeing to import more of our products." China would be importing American values, and as the citizens of China gained power, "the power, not just to dream but to realize their dreams, they will demand a greater say." The minute Clinton said that, China was already winning, because Clinton censored himself. He didn't want to provoke

China. So he didn't say, "This will lead to democracy in China." Instead he just hinted at it.

Clinton was naive to think a little American pop culture was going to mow down 2,500 years of tradition. In the twenty years since, voting has not expanded in China one lick. Globalization isn't bringing democracy to China; it's bringing censorship to the rest of the world. 99% of that censorship is self-censorship.

So today we have three major countries trying to impose their worldview on everyone else. The United States creates pop culture. Russia sows chaos. Both of those are information; they're digital. China's growing dominance comes from its mastery of the physical world. By exporting cities, and the money to build them, it too quietly seeds its digital/information control.

In the years before I came to IndieBio, I used to do a lot of lucid dreaming. It was a bit like Bran Stark in *Game of Thrones.* Except instead of going backward in time, I would go forward in time. I didn't blend into a tree like Bran did, but I'd blend into my bed, and my eyes would roll back in my head.

I would play a kind of chess game, set in the 2030s. In this game, there were five players. Player one was Climate Change. Player two was Artificial Intelligence. Player three was the Genetic Revolution. Player four was China. Player five was the War on Truth. Each of these was, on its own, a history-altering supertrend. That they would all be peaking in power on the same time line and inevitably colliding was frighteningly obvious. There were no rules to the game, except for the premise that I insisted any move be realistic. I was trying to figure out how the world was going to play out.

Every night, I'd dream, and when I woke up, a few pieces would have moved.

I was working as a futurist, and this five-sided chess game seemed to me to be the ultimate challenge for a futurist to solve. Which of the five players was going to impact the world the most? Which would dominate? Which would subjugate? What would the world look like by the end of the 2030s? By day I would study, write, and talk about future technologies and trends. Everything I learned would influence the game I was playing in my predawn dreams.

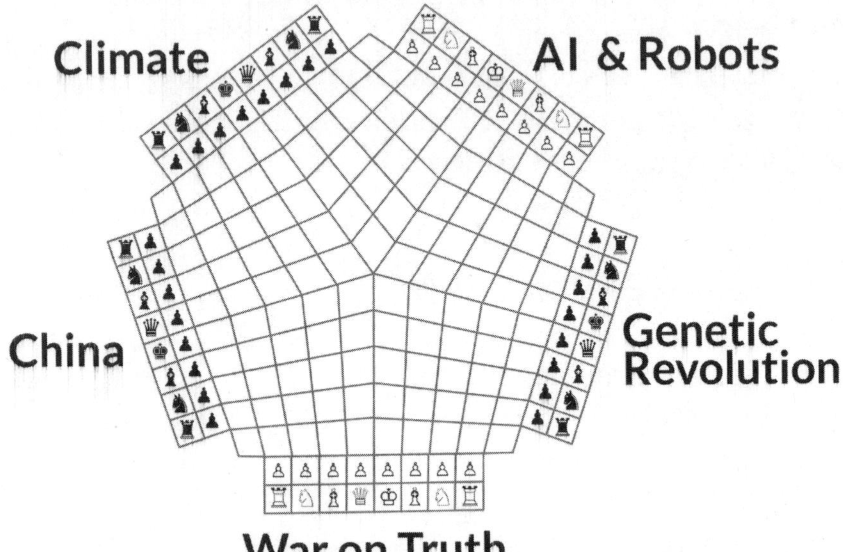

Gradually, it became clear to me that AI creates a winner-take-all economic system, and the result is income inequality. One could say that income inequality is AI's "pollution," analogous to the way carbon dioxide and ocean plastic are the externality created by industry. And fake news (or propaganda) was also a kind of pollution—an idea pollution. The externality created by the genetic revolution could be any number of potential hazards—biosecurity, ecosystem accidents, eugenics—though those were speculative. The climate crisis was going to alter the genomes of all the microbes, which was going to be much more of a problem than rising sea levels.

When I came to IndieBio, I had to stop my lucid dreaming—I didn't have time. But my understanding of the five players increased dramatically. And when my mind wandered back to the five-sided chess game, the future would play out differently. In every run-through of the game, China now had more impact on the world than any other player. Partly because China is secretly controlling all the other players.

Consider that China produces *double* the greenhouse gases of the United

States. They also have three times as much solar power as we do. China invents their own "truth" with a state-run media and will enforce it on two-thirds of the world. China is ahead of the U.S. in using AI–largely because they have national, centralized control of all their data, such as every citizen's health data. In the United States, the data isn't centralized, and health data is protected. China is not equal to the U.S. in genetics, but some things are legal there that aren't here, such as putting human genes in monkeys. Income inequality is now just as bad in China as it is in the U.S., but it's communist with a social welfare system, of sorts. Their rate of poverty is less than 1 percent.

China has the stomach to do things on a national level that few other countries have. When COVID-19 infected Wuhan, the national government shut down the city of 11 million more aggressively than any other country later. They used phone data to grab people out of their houses and bring them in for testing. Drones chased stragglers back indoors, barking orders. That same level of control and obedience is ready to be invoked for other national priorities.

So I believe China is going to change the world more than anything else will. At least in the next fifteen years. But this transformation will largely be invisible to Americans. The only people who'll see it are those who travel to places like Sri Lanka, Croatia, Laos, and Pakistan. And what they see won't *appear* to be particularly noteworthy. They might think, "Wow, this Kenyan train is sure nice." Or, "This is a darn nice bridge Croatia has." It won't be apparent, to casual observers, that this is exactly how China is spreading its system around the world.

But Americans and Europeans will have to truly face the music about what "Made in China" means. "Made in China" used to be a story about *quality*–at first low quality, then suddenly, really high quality. But from 2020 onward, "Made in China" won't be about quality. It will be about ethics. It will be about what we think about when we think about China.

In the fall of 2019, huge American institutions like the NBA, Activision Blizzard, Tiffany, and Vans all self-censored themselves to appease the Chinese government–and were savagely attacked for it back home. We live at a time when China could easily create a deepfake with Houston Rockets star

James Harden saying whatever they wanted him to say—but such trickery wasn't needed, because the real Harden was going to say it anyway.

This was a taste of things to come. Americans—and Europeans—will be asking themselves, "Do I want to buy something made in China? What if I don't like their position on civil rights?" The challenge will be that very little that Americans buy is not, in some way, made in China. China has become so central to global manufacturing and supply chains that even products assembled in America, by Americans, rely on parts and ingredients made in China. The more that corporations withdraw their manufacturing from China, the harder it will be to *sell* in China.

So while we might imagine we'll take a stand, we're actually *already* in deep. Arguably, we're more reliant on China today than any of the 125 countries taking China's trillions for infrastructure ever will be. We saw this play out in Trump's trade war with China. When the United States slapped tariffs on Chinese imports, it didn't help American companies, it hurt them—because their parts and products were all made in China. The U.S. was, ironically, tariffing itself. And then was hit again by China's retaliation.

This exact same dynamic will play out as we try to stick up for the things Bill Clinton hinted at in March 2000. The more we stand up for free speech and try to stick it to China, the more we'll only be sticking it to ourselves. But stick it, still, we will do.

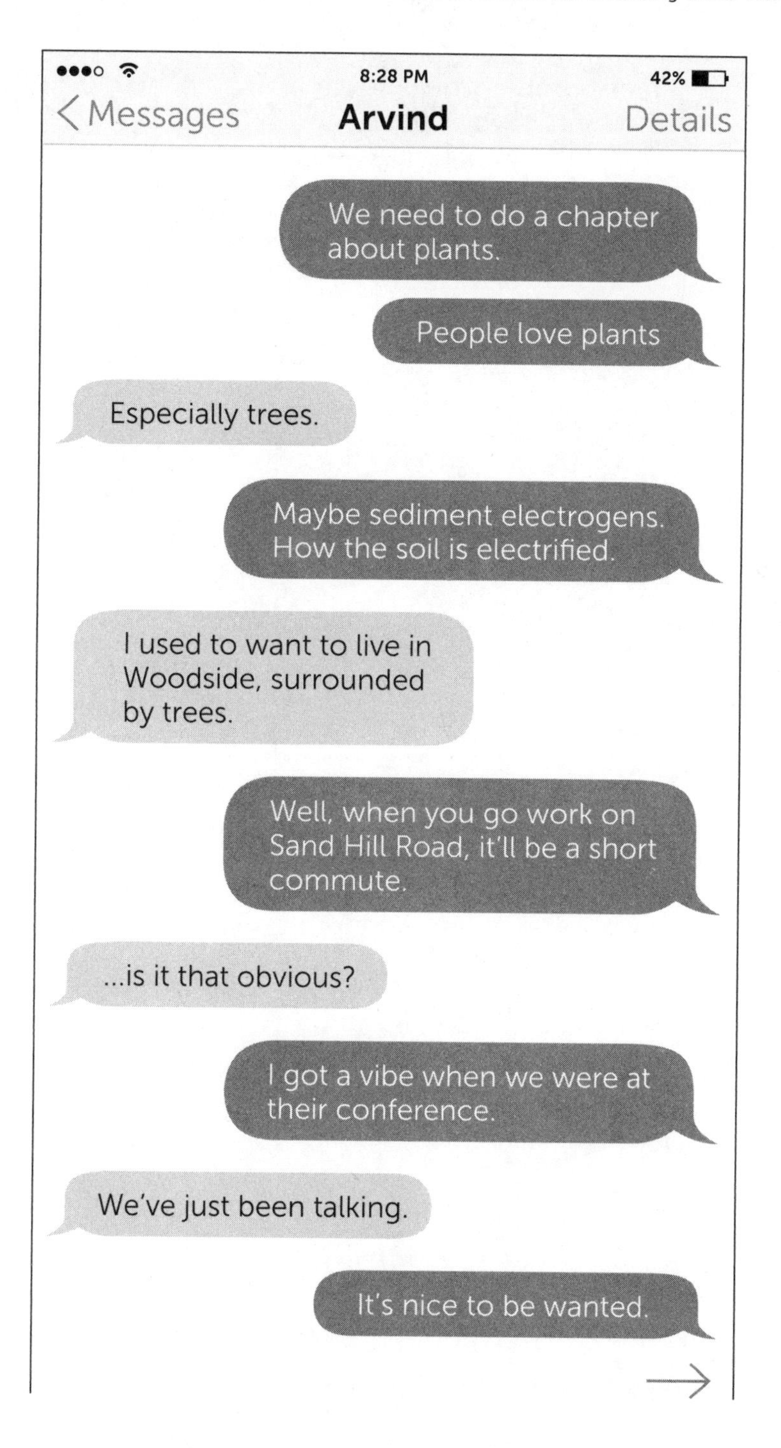
8:28 PM
42%
Messages
Arvind
Details
We need to do a chapter about plants.
People love plants
Especially trees.
Maybe sediment electrogens. How the soil is electrified.
I used to want to live in Woodside, surrounded by trees.
Well, when you go work on Sand Hill Road, it'll be a short commute.
...is it that obvious?
I got a vibe when we were at their conference.
We've just been talking.
It's nice to be wanted.

Yeah. I don't know if I'll do it though. Leave all this?

Would you have more money to deploy?

Way way more.

Text Message

24

How CRISPR Will Change the World Bloomberg

Let's just get this out of the way. CRISPR will not *change the world.* Not anytime soon.

OK? Whew. Like ripping off a Band-Aid. Let's get started.

I am on top of Cerro Frias, a large bump of a hill surrounded by Andean giants. It is just outside El Calafate in Patagonia, Argentina, near the bottom of the world. Sitting in the 4x4 that brought me to the summit, I prepare to get out. The stunning vista sweeps in 360 degrees and would be beautiful if not for the near-constant one-hundred-kilometer-an-hour winds howling from the south. The last gust hits 112 kmh. These winds have circumnavigated the planet and not seen land once until they blast into our Jeep and scour the summit of this hill clean like a giant, God-like leaf blower.

I'm here to experience firsthand the extreme conditions that give rise to novel CRISPR-Cas enzymes.

I push the door open against the unflagging wind. I'm hit full force with the Patagonian gale, and I stagger a bit until I find my footing. I get instant brain freeze from my first mouthful of air. Walking around the barren, windswept rock summit in random circles, I look downward, looking for signs of life before mine runs out. I spy a small perfect crater, like those on the moon, about a palm's width in diameter and as deep as my thumb. Inside I find what I am looking for. Life. A mat of what look like small, succulent green flowers with white petals are pasted to the inside surface of the crater. It looks strange amid the howling wind. I reach down to touch them. They are hard as the stone around them. Lichen, most likely. A symbiosis necessary in this brutal environment. It gives me hope that this little crater has developed its own little ecosystem of fungi, rock, bacteria, and viruses. All competing for light and energy, coevolving weapons against each other, cut off from the rest of Earth. It's in extreme-yet-isolated conditions, such as this barren place, that bacteria's CRISPR systems evolved differently. These CRISPR systems might cut genes differently, or perform the same cut but at a different temperature. Or they might destroy viruses not found anywhere else.

I really should have brought a capture plate and a specimen bag to take samples back to the lab. Instead I take a photo as a consolation prize. My mind drifts to the monumental discovery of CRISPR and the ensuing patent jam. It strikes me that innovation and evolution are one and the same. Nature has invented the greatest gene editor, but it's humans who claim ownership over who gets to use it.

The patent fight between UC Berkeley and the Broad Institute (a collaboration of MIT and Harvard) keeps on going, appeal after appeal. Which really sucks because CRISPR technology is good. It's as if researchers were doing 85 mph on the freeway toward innovation city and suddenly hit bumper-to-bumper traffic with cars filled with lawyers.

Seven years ago, two competing teams between UC Berkeley, led by Jennifer Doudna, on the West Coast, and the Broad Institute, led by Feng Zhang, on the East Coast, discovered the immune system of a bacterial cell and demonstrated its use for gene editing at roughly the same time. Both teams filed patents. Both teams claim they were first to file. Both institutions are fighting to the death over the rights to CRISPR for a simple reason:

cash money. Billions of dollars of licensing royalties are at stake. And the reputations of their hallowed institutions.

This East Coast versus West Coast showdown makes Tupac versus Biggie look small.

Even if the two sides in the patent case settled their dispute, CRISPR wouldn't be available to everyone to use. You may think everyone can use CRISPR and just pay a royalty, but that's not how it works. A small group of companies paid licensing fees to each patent holder for exclusive rights to use CRISPR in certain domains–therapeutics, diagnostics, agriculture, and the like.

This is tantamount to the inventor of the movie projector, Eadweard Muybridge, saying only this handful of people ever gets to make movies.

So much innovation is *not* happening because of the legal battles and restrictions. For example, the small number of owners of the exclusive licenses are busy drawing patent lines around their companies and doing their best to protect their invaluable licenses. They are also raising large amounts of money to fight more court battles that are coming. This distracts them from full product development and getting to market as soon as they could without the fight.

There are only two uses of CRISPR approved in humans. One is for inherited blindness–CRISPR works in the eye because the eye is not defended by an immune system (which anywhere else in the body would attack CRISPR). The second is for spinal muscular atrophy, the leading genetic cause of death for infants. But the treatment damages the liver, as the liver tries to clear the virus from the body.

We have funded a few companies that use CRISPR technology for different applications. Dahlia is a diagnostics company that uses a modified version of Cas9, also invented by the Doudna lab, to glow when it binds to RNA. Dahlia was able to get a license because its founder, Un Kwon, was an executive at Caribou Therapeutics, another Doudna lab CRISPR spinoff. Dahlia is building a research tool to see which genes are active at any given time, and sort the cells based on the level of activity. Dahlia hopes to use this technology to create better cell therapies for cancer and other diseases.

But we also funded another company, Caspr. They're out of Argentina. Unlike Dahlia, they do *not* have a license from the select group of patent holders. Even though one of their advisors, Luciano Marraffini, was one of the early coinventors of CRISPR. Marraffini has stayed away from the patent fights, wanting nothing to do with them.

Without a license, Caspr–our company–has had to look for novel CRISPR mechanisms. Caspr just completed an expedition to a remote area near Salta, an alien landscape pocked with tomato-red pools and boiling mud pots, where the temperature swings forty degrees a day.

During their time at IndieBio, which was just a few months ago, Caspr discovered a brand-new Cas9 variation and a novel Cas12a variant. When coronavirus started killing people in Wuhan, China, it took Caspr only two weeks to develop a rapid, one-hour diagnostic test to detect coronavirus infection. It doesn't diagnose symptoms, like other tests. It detects corona at the DNA-matching level.

So Caspr isn't using CRISPR to edit the genome; it's using CRISPR to detect the presence of a code–in this case, coronavirus–but they've also done Zika and many superbugs. CRISPR's main feature is it cuts the genome when it spots a target code. Cas9 cuts it once, while Cas12a cuts it many times. This cutting feature is used to cleave a bond and release fluorophores, which light up. Much like a test strip for a hot tub, or urine test strips, the change in color is how we know what's happening at a genetic level. The perverse irony for Caspr is that even though they have the patent on their own Cas12a, they don't have the right to *use* it–if it's used in too similar a way to previous patent holders.

The Caspr team is good. Really good. But the machinery of capitalism and patent protection wants to erase them from the earth. *Go away, you innovators! CRISPR belongs to us!*

I flew to Argentina to keynote their national biotech conference. But Argentina is also the perfect place to get a perspective on the patent battle, because of a previous Argentine invention that has dramatically transformed human health.

It was more than forty years ago that another scientist from Argentina,

César Milstein, invented the hybridoma technique to produce monoclonal antibodies. Milstein won the Nobel Prize in Medicine in 1984. Milstein never patented the invention. Instead, he let it out into the world for all researchers to use and develop into therapies and tools for free. César believed that his invention was "mankind's intellectual property." César Milstein would have been a billionaire if he had kept the patent (he died in 2002). Because he made it open source, hundreds of thousands of researchers around the world instantly used hybridomas to create monoclonal antibodies for everything. Forty-four years after he published the hybridoma technique, hybridoma-derived monoclonal antibodies now represent the frontline treatment for the world's biggest diseases, including Keytruda for cancer doing $10 billion a year in revenue and Humira for arthritis with a whopping $20 billion per year. The technique has also been used in countless labs to discover new drug targets and knowledge of biological systems. César Milstein was selfless in his act of prioritizing society over personal wealth, and millions have benefited.

CRISPR is monoclonal antibodies all over again. But this time, there is no César Milstein.

Back in Buenos Aires, I meet with Mariano Mayer, the secretary for entrepreneurs and small and medium enterprises, and a rising star in Argentine politics. Mariano has a youthful face, and bountiful curly dark brown hair with a hint of gray. Slim and tall and elegant, he speaks easily and precisely, as a former patent attorney would.

We immediately begin to discuss Argentine biotechnology and Caspr's potential patents. We have a long conversation about the tension between incentives for inventors and enabling broad innovation from a fundamental discovery. My take is that one company cannot possibly translate discovery into all the applications possible, and society suffers. The minister mentions that the European and American patent systems really differ only in how they are litigated. In the American system, he continues, the patents are issued freely and then argued in court if there is a dispute. In the European system that Argentina follows, he says, patents are ironclad once issued but take years to grant. The trade-off for speed through the patent system is the mess at the heart of CRISPR today.

I sip my coffee, and Mariano sips his maté. We now begin to follow the thread of patent law.

Once a patent is granted, it comes with a twenty-year competition-free protection. "Twenty years is too long for exclusive protection," I find myself saying aloud. Such a long, exclusive window creates the patent fetish that biotechnology and other tech industries revel in. Lawyers rise above the scientists. More energy should be spent in developing great products and winning in the open market. But of course the very costly and risky endeavor of inventing something should be rewarded, or everyone would wait on the sidelines. I continue to think aloud: "What if every patent came with an exclusive for say, seven years, then receive a 5 percent royalty on all sales from other companies using the technology for fifteen more years?" I ask. His face visibly brightens. "That's a compromise that could work," he muses. I think even Caspr would go for it.

Mariano adjusts his long, rectangular black eyeglasses.

CRISPR works nearly perfectly on cells in a petri dish. But in a human guarded by an immune system, we're not there. This is one of the most important lessons you can possibly take away from this book. At IndieBio, we use CRISPR almost every day. Lab work is not under the license restrictions. We edit *cells*. We *program* cells, and then we tweak the program. CRISPR is beautiful. But only at the cellular level.

Inside a human body, our immune system hunts for agents like CRISPR. Our immune system is no joke. It is exceptional at finding anything foreign in the body and attacking it–often so violently that the defense response is worse than the disease. There are proteases and nucleases that cut things into little pieces. And even if we design ways to sneak past those to successfully deliver a gene, we can never do it again. The immune system uses T cells and antibodies to remember enemies.

Even if we get CRISPR past the immune system to the cells in our body, there are off-target effects. The guide RNA is eighteen to twenty base pairs long. That *should* be long enough to never accidentally match a random-but-identical base pair pattern in a human genome with 3 billion base pairs. However, in practice the base pairs are sticky, like the suckers on an octopus arm. They grab across and around each other to different base pairs, creating a false fit. Like shoving a Tetris block into the wrong hole. Then activating the Cas9 cut and resulting in toxic or fatal side effects as it turns off random genes in the patient's genome.

At IndieBio, we are investing heavily in ways to sneak CRISPR past the immune system, and to avoid these off-target effects.

But even with these massive problems to solve, before it can be used on humans, CRISPR is widely regarded as being worth hundreds of billions of dollars.

My next meeting adds a different voice to the conversation.

I arrive at a gorgeous building in the swankiest part of Buenos Aires. The building is a national landmark. Inside, the lobby is studded with paintings and sculptures by Anselm Kiefer, along with other modern masters. It looks more like a marble-walled museum than an office. I am at the headquarters of Grupo Insud, a leading global manufacturer of drugs and biosimilars. I meet the founders in their splendid solarium for lunch. Hugo Sigman is an older gentleman with a very soft voice and a gentle smile. His wife, Silvia Gold, is sparkling and sharp, with a PhD in biochemistry. Together the two founded Grupo Insud in their twenties and have passed on the daily operations to their son, who is now CEO. Hugo is the president of the Argentine Chamber of Biotechnology. We take our seats, and the first course of egg terrine arrives.

We turn our attention to Grupo Insud's core business of biosimilar and generic drug production. They are a leading manufacturer of antibody therapies that have gone off patent, including Rituximab, a treatment for lymphoma and arthritis. They represent the other side of the patent cliff—when it expires. Grupo Insud can manufacture the same drug and offer it at a much lower cost but still make plenty of money by eating into the sky-high margins the patent protection offered the patent holders.

"The true spirit of the patent's profit protection is being broken by the big pharma companies," said Hugo. Even though the biologics come off patent, the pharma legal teams find ways of reapplying for patent extensions and other shenanigans to keep us from making the same thing at lower cost. "The real loser is the patient and the public health care system," Hugo finishes.

I ask about the twenty years of patent protection and if a shorter cliff would be better. "Of course shorter cliffs would help my business, but I am in the innovation business, too," Hugo says. It's a division of Grupo Insud called mAbxience, which is developing novel cancer therapies, and has one advanced program. He is looking for venture capital to help shoulder the risk of development. The high cost and risk of development requires him to have the patent protection, and the more the better. If you can't beat 'em, join 'em.

Two weeks later, I'm back in San Francisco when Franco Goytia, the CEO of Caspr, walks in. Hugs all around. Just two days ago, Caspr was in the news all over Latin America for inventing its DNA-based detection of coronavirus. As of this morning, coronavirus has killed 1,018 people. Now Franco has something more to show us. Our IndieBio team piles into a room, and Franco gets his team on the phone from Argentina.

"We found a way to get our own method-of-use CRISPR patent," Franco announces.

"Walk us through it," I ask.

"To be patentable, it has to both be different and improve performance," Franco says. "We're going to use our own Cas12a, but that alone isn't enough. We're also using it a different way: pairing it with DNA Origami."

There's a pause as we all take this in, thinking about how DNA Origami could apply here. DNA Origami is a way of controlling the shape of DNA with genetic keys. These genetic keys can also make a single-stranded piece of DNA self-replicating.

Even before Franco gives us the details, we realize this *could* work.

"When we detect the virus genome, the cut isn't releasing a fluorophore, like the prior patent. Instead, the cut triggers DNA Origami to begin," Franco explains. "The self-replication process kicks off, multiplying the signal hundreds of times. Then it releases the fluorophore signals. So we get a much better signal."

"What's your data look like?" I ask.

"It's a considerable boost in performance," Franco responds. That means it's patentable.

Franco clarifies that they *may* still need a license because they're still building on the foundation of CRISPR. But now that Caspr has its own patent–and a better solution–it has more leverage at the bargaining table.

I couldn't be more proud of our company. Not just for surviving in the harsh landscape of patent law, which tries to suffocate the life out of small players. But for fighting with science, not lawyers.

25

The "Blood Boy" Clinic Is Coming to NYC So Rich People Can Live Forever Mashable

About five or six years ago, the salon crowd of Silicon Valley would gather, and for the evening discuss the beguiling work of a dozen scientific giants in the new realm of "longevity." If they were lucky, Liz Blackburn would be there in person to explain telomeres and how stress was literally shortening our life spans. Or Tom Rando might be there, to talk about the old mice he hooked up to young mice, which made the old mice grow younger–like reversing aging at high speed. Eric Verdin would describe how different compounds made mice live 40 percent longer. Real money had started to flow into companies to capitalize on this body of work. Google had a new secret company devoted to the field. If we could cure aging, we could cure all disease.

These salons were always quite serious and awe-inspiring. It also felt like being let in on a secret. Not a secret kept by humans, but a secret kept by life's code: Aging was hackable.

So here we are now, it's 2020, and I feel safe assuming that almost everyone has heard *something* about the longevity movement. The space has exploded. There are supplements you can buy. Companies have gone public on the promise of their magic. Thousands of compounds and pathways are being studied. Blood boy transfusions have been mocked on the show *Silicon Valley*. We have two longevity companies–one that reverses aging in skin, and one that gives your cells more energy. Studies of mice jumping on their treadmills again are everywhere. But at the same time, nobody is yet living longer. No human is growing younger, like Rando's famous mice.

So what are we to think about this whole field?

We actually love this sector. But everyone needs some context.

First of all, humans *already* live a really long time. Comparing across species, we are way up there in our life span. You might say, we're *already* quite optimized. We're *already* hacked. We live for many decades beyond our reproductive years (and we count both genders in that, because the older men are when they father a child, the more mutations they're passing along). The main physiological way we live so long–the main way we stretch the end of life into decades–is by slowing down our metabolism. This is how Galapagos tortoises live for 150 years; they slow it all down. Except they live their entire life in slow mode. We humans only do it after age fifty or sixty. But when we shift into lower metabolism, we might go to 95, or 110, if we are otherwise disease free.

Mice, meanwhile, do not live a long time. About two years. They are far from optimized for long life spans. This makes them pretty easy to hack. There's about a hundred ways to make mice live longer, but they don't necessarily translate to humans. A mouse has a heartbeat of 350 to 800 beats per minute. Its metabolism is insanely fast. An average mouse weighs 25 grams. It has to eat one-fifth its body weight every day because its metabolism is so high. That would be like a 200-pound human who eats forty pounds of food a day. Some of this is just thermodynamics. Smaller animals have a higher surface-to-volume ratio. They lose heat to the environment much faster, so they have to generate more heat through eating.

So mice are not great models for human aging.

But a bat is around the same size as a mouse.

And bats are even better than humans at living long. Bats are longevity-optimized already. Bats can live forty-five years or more. They have the same size-related thermodynamic challenges as mice, but not the short life span of mice. Bats and humans both learned to live in caves, or under shelters. Also, the bat genome includes more DNA repair genes; if you recall from our cancer chapter, these are critical to keeping your genome from being torn apart over time by free radicals. These mechanisms of rapid DNA repair also make them less vulnerable to viruses.

So you might ask yourself, "Have I ever read a longevity study where they made bats live longer?" Nope. Admittedly, it's tricky; bats live too long to get the results. But my point is, the bat is an organism–like the human–that lives a long time. A bat study would be far more impressive than a mice study.

Most older people would say, "Oh, I'd love to have a faster metabolism. I'd have more energy, and I would enjoy food more and not worry about my weight." The promise of the longevity field is that eighty would be the new fifty; that all these septuagenarians would be out playing rugby, or crushing it at the office. A faster metabolism would certainly make you *feel* younger, but nobody should leap to the idea that it will make you *live* longer. It might end your life sooner. Because a slowed metabolism is a big way we survive so long now.

Now let's understand this a little better, because it's not quite what it first seems. We have a strong tendency to associate late-life vigor with physical activity. And we also associate slow or fast metabolisms with weight gains and slimness. But set that aside for a moment, and think about what's really using energy in the body.

The brain is a much bigger factor in life span and longevity than we realize. And the part of the brain that really matters here is the cerebral cortex. Human brains have 16 billion cortical neurons. Unlike the other cells in our body, these aren't replaced. They last a lifetime, and so it starts to make sense why they might play a big part in determining life span. If you can keep your brain alive, you can probably keep the body alive. One of my favorite studies came out of Vanderbilt in 2019. It looked at the brains of more than 700 species of mammals and birds (the warm-blooded animals), and specifically at the number of cortical neurons in the cerebral cortex of each species. That number *very strongly* predicts the species' life span. So the intelligence we recognize in parrots explains why they can live sixty years.

Exactly why having more cognitive complexity makes a species live longer isn't totally decided by science yet. But the simplest answer is probably the best answer: The species is using its intelligence to survive. Smart animals figure things out. They solve problems and learn to avoid predators.

So the number one reason that mice have short lives and humans have long lives is not our different speeds of metabolism. That's number two on the list. Number one is that humans are much smarter than mice. This makes mouse studies somewhat irrelevant to human longevity. In fact, if you *really* wanted to extend human life span, you'd design humans to have even bigger brains. We would think our way to longer lives.

One of the most important genes for human longevity is a brain gene. The name given to this gene is onomatopoeic. It's nicknamed REST, which kind of implies its purpose–it puts brain genes to sleep. Its real name is RE1-Silencing Transcription Factor. To explain the role of REST, first recall that the entire human genome is in the nucleus of every cell. All 3 billion base pairs. And what this means is that cells throughout the body carry brain-related genes in them. You really don't want your kneecap expressing brain genes, or kidneys acting like a brain. So your body needs to silence these brain genes in cells outside the brain. It uses REST to do this. REST encodes a protein that sits on your DNA, recruits some other enzymes, and tells brain genes to sleep.

People with more active REST genes live way longer. Almost every human who lived more than ninety years with their mind still sharp had a very *active* REST gene.

So if I was introducing REST at an awards dinner–maybe a "Genes of the Year Awards"–*The Genies*–I'd say, "REST is the gene that makes your brain be a brain, and makes your body *not* try to act like a brain."

REST is a big deal. But you won't find it on a 23andMe test. Because it's not like some people have it and others don't. We *all* have REST. We all *need* REST. What matters is how *active* REST is. Most people with Alzheimer's disease have suppressed REST genes.

REST is active in the brain, too. It makes the brain more efficient. It's a neuron silencer. Scientists can measure how much energy the brain is using. REST gets it done with less energy. And so this really helps longevity. People whose brains use less energy live longer. REST is more active in them. Recall for a moment our learnings from the chapter on the origin of life: Evolution favors energy efficiency. More efficient brains get to live longer.

So let's summarize for a moment, before this story turns.

1. In warm-blooded animals, the number one factor in longevity is cognitive capacity.
2. After that, it's metabolism.
3. To live long, you don't want a fast metabolism, you want a slow one.
4. Our brains use a ton of energy, and people with more efficient brains will live much longer.
5. Be wary of extrapolating from mice studies to human longevity.

The brain is the most important and most exciting vector of longevity. But it's also the least studied. It's really tricky to study the brain. A scientist can study human neural cells–a vial of a million cells costs $570 online at Thermo Fisher–but these are neural *stem* cells. Fundamentally, baby cells. When what we really care about is what happens to those cells when they're seventy years old and beyond. Remember, these cells in our brain *have to last a lifetime*. But scientists can't wait that long. Even when they test human neural cells, they let them divide only three to eight times. So even human neural cells aren't a great model for what's really happening in our brains.

So we're back to studying mouse cortical neurons ($425 for a million). The mouse neurons are also embryonic, taken from fetal mice seventeen days after conception, but in only a year or two they're "old" and can be studied. The main reason scientists use mice studies for longevity is that mice age fast, and so the research can be performed in a couple years. They also use worms–their life span is just a few weeks.

Now I need to draw an imaginary line in this chapter. Actually, let me draw a nonimaginary line.

Everything above that line is declarative. It's our opinion but it's a strong opinion. We can make those few blanket statements.

But below that line, we can't make any blanket statements. Nobody should. Below that line, it's a minefield of trade-offs and compromises. There are many inherent dangers in longevity science. Potential unintended side effects. Scientists searching for longevity compounds have to carefully navigate the terrain, finding ways to get the benefits without the risks.

So imagine for a moment it's the future, and two pills have been invented. The first pill works as dramatically on older humans as Tom Rando's blood sharing did on older mice. In just a month on the pill, you shed a decade off your life. Everyone who has taken it feels great, but nobody has taken it for more than two years. The other pill works slowly. So slow that you really can't even notice the effect. You still feel old. Months go by, and you're not really sure it does anything, even if your doctor says, "It's working."

Which pill would you take?

I would suggest taking the second one.

I don't know if we will ever see a pill like the first, but if we do, be wary. I would much rather be on a pill that, before anything goes too wrong, a doctor could see the problem in my biomarkers and get me off the pill.

The most fundamental trade-off that longevity science faces is that many of the mechanisms of longevity–at least in cells–are also the mechanisms of cancer. Cancer cells use a series of genetic hacks to make themselves immortal. Cancer cells taught science a *lot* about longevity. So longevity researchers need to figure out how to induce longevity *without also causing cancer.*

In fact, I'd say they have to do better than that. Because, as Arvind wrote in the cancer chapter, the number one reason humans get cancer is because we live a long time. Every time a cell divides, it's another pull on the cancer slot machine. Living longer would mean even more chances to get cancer. So to succeed, longevity scientists have to figure out how to induce longevity *while simultaneously suppressing cancer.*

We can see this hair-splitting challenge in many cellular mechanisms that have been studied. From lengthening telomeres, to cleaning the junk out of cells, to DNA repair, to entering the zombie state of senescence. The hallmarks of longevity share quite a few features with the hallmarks of cancer. Cancers are also diverse, so in some cancers a longevity hack will suppress cancer, but in other cancers that same hack is used by the cancer to survive.

So intuitively, it makes sense to be most excited by longevity compounds that simultaneously show a suppressive effect on cancer. To illustrate this, let's use two compounds. One is metformin, the drug that is currently getting the most hype among longevity advocates. The second is MitoNova, which one of our companies developed. They share some similarities. *As I describe these two compounds, please remember we are officially "below the line." And we are not giving medical advice.*

Metformin has been used since the late 1950s as a type-2 diabetes drug. It's the synthetic version of the active compound in the pale violet flowers of the French lilac bush, which was used as an herbal remedy for centuries. Legend has it that goats won't eat French lilac because it lowers their sugar levels too much.

So why would a cheap, old diabetes drug also be considered a longevity treatment? Think about what you've already read in this chapter; the clues are there. Diabetes is a disease where our bodies lose their ability to efficiently balance sugar levels in the blood; sugar is a source of energy. So one could describe diabetes as the inability to be energy efficient. Metformin is a remedy for that. It makes our bodies more energy efficient. It improves our metabolism by shifting our energy over to burning fat. Life, and evolution, favor energy efficiency.

In the brain, metformin seems to have neuroprotective properties. Not every study shows this; and all the studies are only on patients with type-2 diabetes. And while we're wary of mouse studies, they do show metformin increases mouse life span.

Simultaneously, metformin has been shown to be an anticancer compound. Studies of people taking metformin show a reduced rate of cancer and reduced progression of cancer. These are type-2 diabetes patients, but many oncologists are now giving metformin to cancer patients alongside chemotherapy, and the results have been favorable.

At this point, metformin starts to sound like a wonder drug for the aged. It's got the best of both worlds. But before getting too excited, remember we're below the line. Trade-offs come with the territory. For metformin, the trade-off seems to be that *metformin counteracts the benefits of exercise.* Exercise is wonderful for the body. Running will add six years to your life. The moderate stress on the body created by exercise signals all sorts of regenerative processes that improve health.

The very same people who might love to get their hands on a bottle of metformin are already on the treadmill, maintaining their physical fitness beyond the age of sixty. But metformin will throw a wrench in the exercise path to staying young. One might say it competes with exercise. The researchers who did this work took muscle biopsies from the participants' quadriceps, and they found that the muscle tissue mitochondria were compromised by the metformin. They estimated the negative impact at 40 to 60 percent. It didn't harm the subjects, but it meant the benefits of exercise were greatly reduced. So the team of scientists who did the study were definitely worried about a future world of older people, all taking metformin. Their paper sent up a big warning flag.

It's not clear at all that metformin lives up to the grandiosity of being a longevity drug. It certainly doesn't make you rapidly grow younger. Most people who are on metformin barely notice it. Perhaps it's just a good drug.

MitoNova doesn't have a century of research behind it–only about a decade of work. But it comes from something that has millennia of evolution behind it: mother's milk. The compound is a peptide-mineral complex that babies naturally make, in their gut, from two ingredients in mother's milk–zinc and alpha-lactalbumin.

Mother's milk is truly underrated. Even while pregnant, mothers begin creating the compounds of milk, and these are found in moms' circulation and in the amniotic sac of the embryonic baby. Once the baby is born, mother's milk replaces the placenta as the main way a mother's body communicates with an infant's immune system. There are hundreds of compounds in mother's milk; alpha-lactalbumin is one of them. In a baby's stomach, the alpha-lactalbumin is chopped up by digestive enzymes into chunks, which are consumed as food. But some chunks are the right length to wrap around a zinc molecule and become a bioactive compound, like a drug.

One might assume a couple ingredients in milk couldn't be as powerful as a drug invented by scientists from scratch. I would argue the other way around: Evolution has perfected the compounds in milk over millions of years to be just what the body needed.

Dr. Helen Chen specialized in embryogenesis. While others in her field focused on the development of stem cells in utero, Helen was interested in the growth factors that embryos develop in. It took her years to isolate and understand the role of natural MitoNova, and then figure out how to make it synthetically.

MitoNova ticks all the boxes for a safe antiaging compound. It prevents cellular senescence. It inhibits free radicals from tearing up cells. In the brain, it increases the release of an antioxidant enzyme, superoxide dismutase, and it reduces tau proteins. In cells, it increases ATP energy output by 25 percent. In mice studies, the mice get on the running wheel far more, and they sleep better. In preliminary research on human cancer patients, as well as on dogs, MitoNova has shown to be effective against the muscle-wasting effects of many cancers.

It does all this by acting on our cells' mitochondria. Our cells don't just have one mitochondrion in them; neurons have up to two thousand mitochondria, and heart muscle cells up to five thousand. It's in the mitochondria that we split oxygen and create the reaction chain of energy production. But as we age, mitochondria get leaky, spilling out free radicals. MitoNova uses its zinc ions to localize specifically to leaky mitochondria, and then it acts on the mitochondria's genetics. Four genes are particularly affected. CASP3 and XIAP help protect the cell, so the body doesn't kill the cell. ATG5 and BECN1 both promote killing off the leaky mitochondria, junking them into bits and clearing them from the cell. New, healthy mitochondria grow in their place, and cell function is restored.

Helen's work gives her an interesting insight into the association between metabolism and aging and exercise. "Conventional thinking is we need to be careful about speeding up our metabolism as we age. But that's because our mitochondria are leaky. If you speed up our metabolism, you just crank out more free radicals into the cell. And those cause all sorts of cellular damage. But if you declutter the leaky mitochondria, and let the body regrow healthy mitochondria, then you can safely increase your metabolism."

So unlike metformin, MitoNova might be a great companion to exercise in the aging population. One of its most noticeable benefits is that older people don't get as sore and stiff from exercise. (I took it for one month, and I really noticed this difference.) I asked Helen to explain why this happened. She explained that even when a cell has a lot of leaky mitochondria in it, it has to generate a certain amount of energy. So in older people, the cell is switching from aerobic energy creation over to anaerobic glycolysis, just to get the job done. No matter how young you are, super-intense anaerobic exercise will make you sore. But in older people, even mild exercise will tax cells and switch them over to anaerobic energy. They might have just played a little tennis, but the next day they feel like they ran a marathon.

"By cleaning up the leaky mitochondria, the cells in older athletes now can meet the energy demands with aerobic energy creation again," Helen explains.

In other words, MitoNova doesn't help the older athlete *recover* from soreness. It prevents them from getting sore in the first place. Like metformin, MitoNova makes our bodies more energy efficient.

Yeah, I drew one more line. Somehow, I'm unsatisfied with how this chapter is turning out, and I want to correct it. I want to repair it.

Metformin and MitoNova may be good drugs to help people as they age. But I don't think they're truly longevity drugs. The hype and allure of longevity creates a kind of halo around certain compounds today. The idea of longevity has become this twinkling seam in our culture. It's more of a thought exercise than a real thing. Hundreds of articles have been written about how society will be different if everyone lives to 120. And why not? It's more fun to imagine living to 120 than to confront our own death.

But let's speculate for a moment how scientists might one day *actually* reverse aging. I don't think it will be through a magic pill. It's going to be by acting directly on genetics. There are different camps, or ideologies, in the longevity field, and Arvind and I are in the camp that subscribes to "the DNA damage theory of aging." Which holds that aging happens because unrepaired DNA damage accumulates over time. Our DNA gets damaged and our genes get turned off more and more with time. So "the code of life" within us is still there; it's just not operating like it used to.

Think of a house with several bedrooms and a reading den. The overhead light in the den is glitchy; the light fluctuates and sometimes doesn't even come on. So you learn not to turn the light switch on at all, and use a lamp–or spend less time in the den. That's aging. A failure to repair.

I believe the first human miracle of antiaging that we'll see–and it might be quite soon–is gene therapy for children with a very rare disease called progeria. A single mutation causes the condition, and the condition causes children to age extremely rapidly, rarely living past age twenty. It affects one out of 8 million children. Monogenic diseases can be treated with gene therapies, and the experiments (in rats) for progeria have shown it can be helpful.

Aging in normal people is *not* monogenic. It's probably the most polygenic disorder of them all. To the extent we one day try to treat it, it won't be by fixing all the errors we've accumulated. It'll be by fixing some of the genes that encode DNA repair proteins–so they catch the errors in the first place.

The FDA doesn't consider aging a disease, but there are a lot of diseases that could be treated, or managed, using enhanced DNA repair. Therapies will be invented to treat the diseases. Living longer would just be a side effect.

Until then, living longer remains a thrilling idea, an inspiration. And the mere idea of longevity–the hope this field provides–may alter how we live the days we have.

26

The Meaning of Life: Albert Camus on Faith, Suicide, and Absurdity

Big Think

Inertia does not just operate on the global economy. Inertia is everywhere in our lives, especially in the journey we take through life. You're born, you go to school...

We are taught to question this script. And yet those who teach us to be questioners do not prepare us for the incredible difficulty that can bring. For some, they feel like an outcast. Others can handle being outcasts, but they face an even bigger problem: the futility of fighting inertia.

This sense of futility, this powerlessness, is at the root of so much self-destructive behavior in our society. It's the desire for change, paired with the inability to enact change.

In the last decade, society has seen a dramatic rise in overdose and suicide deaths both in older men and in teenagers. But for every one sudden end, there are ten more, or a hundred more, acting out with less severe manifestations of self-destruction. The script of life created expectations that, in a fast, blind, and dense world, they could never meet. If they questioned that script, they had it even worse, dreaming of changes that could never materialize.

I know these feelings, because I was a questioner.

Being a questioner is scary. It's hard. And once you start questioning, sometimes you don't know when to stop. How deep is too deep?

Starting around when I was twenty-six years old, I put my family through several years of pain. I had torn my life apart. I had quit a job that was paying me a lot of money, and I had given up my dream of becoming a physician or researcher. Instead, I had no job. Deliberately. I was living in Berkeley, on Regent Street, in a one-bedroom apartment–with five other people. The cops came by all the time, telling us to keep it down. I slept on the floor. I let a homeless guy, Prince Charles, sleep there, too. He brought in his girlfriend. When I woke, I would scuffle across the street to the sidewalk tables at La Méditerranée, where I would smoke cigarettes and watch Inertia pass me by, every morning, on their way to work. They'd bustle past me in their Patagonia fleeces and JanSport backpacks, headed to BART and the City. I was jealous they had somewhere to go. I was jealous of the sense of purpose in their eager stride. But follow them, I could not.

You do this sort of thing on instinct. It doesn't make sense to you as it's happening. I knew I was questioning the role of money in our society. Not quite the value of money, but the way money carried meaning beyond the amount of a purchase, the mass hallucination that money is virtuous, something to be respected. And to get a perspective on that, I was living without any money. I was looking for that sense of liberation you get when you learn you can live on nothing.

But there was far more going on. Only with decades in the rearview mirror do I really grasp what I was doing. I was taking life down to the bare minimum, down to the studs, down to the rails. I loved my family and I had great respect for the sacrifices my parents made, as well as huge admiration for how they'd come to the United States and remade their lives. But in my twenty-six-year-old mind, everything they had done for me was both a gift and a crutch. I had a deep, individualist desire to accept zero help from anyone, and to wake up knowing my life was made/created by nobody but me. What I craved was raw and primal: true self-determinism. Even if I wallowed in it. "Selfish," my sister called me.

Midday, I would wander over to Moe's bookstore and read the Great Works and write bad verse for the poetry slam events in Oakland. In the afternoon, I was learning to paint from an artist named Masha Savitz, whose work was considered "painted poems." Oh, strident youth. Everything I did was bad. My poetry was cringeworthy, my paintings could be mistaken for elementary school art, and my prose was wooden. But at least it was *mine*. Nobody had made it but me. I started with white paper and blank canvas.

At night, I jumped off buildings and bridges and towers. I'd become a BASE jumper. You do it at night to not get caught. I did more than 260 jumps in those three years. Most of the jumpers I knew back then are now dead. My parents learned what I was doing. My dad would wake up at night screaming.

As I watched the money of Silicon Valley pour into San Francisco, I could see the new towers on Rincon Hill shoot into the sky. When Bridgeview Apartments went up, twenty-four stories tall, I jumped off its roof. This was probably the most inexplicable piece of my life. The goal of BASE jumping is not to jump off the tallest thing you can. It's to jump off the *shortest* thing you can, and still walk away. The Bullards Bar Dam is 645 feet. It's magnificent. Perrine Bridge over the Snake River is 486 feet. Bixby Bridge in Big Sur is 260 feet. The Essex Condos on Lake Merritt were 220. I did them all. I didn't know where to stop.

I was questioning life's design. The code of life. The rules of society. The value of the academic degrees we earn. If family matters more than friends. I was asking if life itself mattered. The code of life does not start with genetics, and it does not end with genetics. The real code is in how society organizes and gives meaning to life. Albert Camus said that giving meaning to life is futile, that it will always escape the meaning we ascribe to it.

I was challenging that.

We're born, we go to school... life is given to us. We don't choose it. It's chosen for us. This entire equation changes when you jump from twenty-four stories. The world is hyperrealistic in that moment. Peaceful. Silent. You feel the turbulence of your body through the air as you accelerate. Lights twinkle and dance off the water on the bay. The traffic from the bridge drones a dull brown noise. Then you must choose life, and reach for the pilot chute.

To all who knew me, it appeared I was destroying my life. Maybe. But I never felt more alive. I was learning to choose life. Over 260 times, I chose life, when if I waited a fraction of a second to decide, my life would be taken. My muscle for choosing life got stronger and stronger.

You can say there are 25,000 genes in the human genome. Or you can say it this way: You have a little more than 25,000 days in a lifetime. 25,000 chances to choose life. To be alive. To be present. To make it yours. To leave a mark on the blank canvas that is our journey. There is no gene for it. Life isn't scripted. It's just you.

I walked away after three years. I'd learned what I needed. I had ripped my life down to the rails, and I walked away with the muscle to choose life, every day, ever after. To not tire of the futility. The blank canvas was in front of me. It still is.

For each of us, this journey is unique. In Camus's *The Stranger*, Meursault comes to understand that the meaning we give to life may be futile, but it's all we have.

With the benefit of eighty years of perspective, I can look back and see that, in his own way, Camus was editing the code of life. He was changing the way we think. He was redefining how we infuse life with meaning. To me, the lesson of his work is not that we must learn to accept life. It's that we get to decide what life means. It's up to us. We're born, we go to school... the canvas may not be blank, but it's not finished, either.

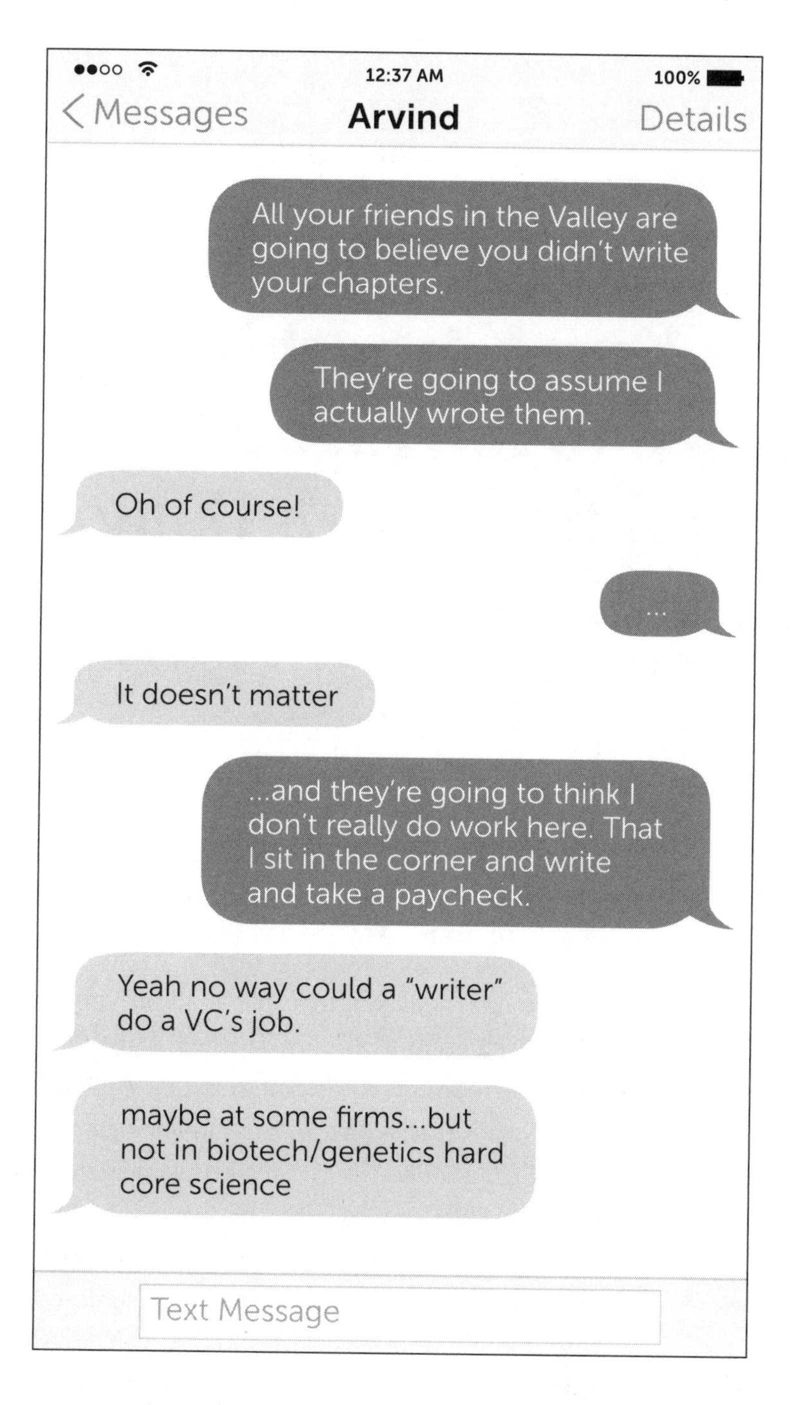
12:37 AM
100%
Messages
Arvind
Details
All your friends in the Valley are going to believe you didn't write your chapters.
They're going to assume I actually wrote them.
Oh of course!
...
It doesn't matter
...and they're going to think I don't really do work here. That I sit in the corner and write and take a paycheck.
Yeah no way could a "writer" do a VC's job.
maybe at some firms...but not in biotech/genetics hard core science
Text Message

27

The Emotional Reason Why Kim Kardashian Turned to Surrogacy (Again) for Fourth Baby on the Way People

I am sitting in an unremarkable nail salon in a strip mall on Ventura Boulevard in Sherman Oaks, just a couple miles from where I grew up. On the low table in the middle of the room sits a pile of magazines, and on the top is *People* with the headline "Kim Kardashian's Emotional Surrogacy." It occurs to me Kim Kardashian is simultaneously the most misunderstood person in the world and the most interesting idea in the world. Her stepmother is Caitlyn Jenner, former American sports hero and Olympics decathlete, a trans woman. She is married to Kanye West, who is diagnosed bipolar and, depending on whom you ask, a musical genius. Kim Kardashian has 151 million followers on Instagram, half the country. She recently had her second child with a surrogate mother, and the news is on fire about it. Like Minute Maid, Kim's life is made from concentrate.

Sitting next to me, my wife is getting her nails done. We are considering having a third child even though we are getting old. I'm forty-five and Krissa is thirty-nine. It's risky. So I mention the Kim Kardashian headline to Krissa and ask her if we should consider a surrogate or in vitro fertilization. She says no, we don't need to. After a moment, I agree with her and add a throwaway, "But if we did, we could pick the gender of the baby and check its health." And that's the moment a millennial with frizzy hair like Einstein, colored neon pink–sitting in the chair across from me–says, "Dude, that's a scary thought."

I know what she means. I hear it all the time.

Headlines every day shout a future where childbearing and age become decoupled, where we can scan and choose from the genetic codes of our embryos, where we can design our babies from a restaurant menu and can bend gender to our will. It freaks people out.

They're worried about this future. Worried about a new race of people chosen by hair color and intelligence, worried that the only fairness of life left–the randomness of our genes at birth–can be stolen by a scientist and CRISPR.

But I'm not as worried about it, and there's a reason why. I think that story line–"designer baby" clickbait–is misreading history and the trend line that projects into the future. Since the dawn of time, fertility technology has gone hand in hand with women's freedom and economic self-determinism. I see that trend *only* continuing. Fertility technology is not going to lead to designer babies, it's going to lead to more equality for women.

The first reproductive technology we know of goes back to an ancient city in North Africa, Cyrene. A plant named silphium was discovered to prevent pregnancy. Women brewed it into a tea. It raised estrogen levels. Silphium tea was so popular that it was consumed to extinction by 100 AD. I don't know what life was like for women in ancient Cyrene. Probably not good. But probably better than in ancient Rome, where the average woman bore twelve children.

Nearly two millennia later, American soldiers came back from Europe with condoms, and for the first time a judge overturned laws against selling them. The very next year, American women won the right to vote. American women got the right to choose their politicians at the same time they got the ability to choose motherhood.

Then the pill powered the '60s, the women's liberation movement, and the surge toward the Equal Rights Amendment. Along with it came, increasingly, economic power. The proportion of women in the workforce climbed rapidly, and every reproductive technology that came along continued to drive workforce participation higher, from IVF in 1978, to using donated eggs in 1983, and cryopreservation of eggs in the 1990s. By the late 1990s, the number of women in the workforce had doubled. Fifty years ago, 7 percent of teen girls had a baby. Today it's 2 percent, the lowest ever. Women earn twice as much annual income in their late thirties compared to their early twenties; having the power to delay childbirth allows them to earn way more.

I explain my thesis to my wife and the woman with neon Einstein hair. A brunette woman wearing a light blue plaid suit speaks up.

"Sorry, I couldn't help but overhear your conversation," she says, taking off her gold-rimmed glasses. "I think it's crazy we have to find a great job, great guy, and have a kid–all before turning thirty-five. Makes me wanna get a sperm donor and take the heat off." She pauses. "Or just not have kids."

We have not reached economic equality–not by a long shot–but that's precisely the point. In Silicon Valley, all the top tech companies offer egg freezing, multiple rounds of IVF, and genetic screening as a perk of employment. The irony is that we are now spending our whole youth trying to not get pregnant, then having trouble trying to do so later in life.

Fertility technology will be a critical driver of improving economic equality in the future.

I pick up the *People* magazine and turn to the story. As I start to read, I blurt out, "Kim and Kanye didn't choose the gender of the baby."

"I thought they picked a girl," said Einstein Hair.

"I thought so, too," I said. "Everyone says that. But it's right here in

People. They just wanted a *healthy* baby." The doctor picked the embryo with the healthiest characteristics. That embryo just happened to be a girl.

The future we are sold is a fertility clinic that will one day bend genders with same-sex parents, three-parent babies, screening thousands of embryos derived from your own skin cells and editing the chosen embryo to your exact desires. No wonder they are freaked out. This future isn't human.

It's also not possible.

To understand why, we need some basics of how we grow. When a sperm fertilizes an egg, it starts a series of chemical reactions that kick off cell divisions. The important thing to keep in mind here is that just one cell turns into red blood cells, white blood cells, nerve cells, and over two hundred different cell types, all before birth, in a process scientists call differentiation. You can think of it like a tree, with the embryo at the trunk. At each branch is a stem cell that has the ability to become other cells. After five days the dividing oocyte becomes what we call a blastocyst, a ball of about 256 cells that hatch out of a protective outer layer, much like a chick from an egg, and implants in the uterine wall. On day fifteen a milestone appears as a visible streak across the fetus. Called the primitive streak, it is the beginning of the digestive system and the brain. It is also the moment when regulators have drawn a line in the sand; no further experimentation on embryos is allowed from then on.

When Kim Kardashian or the thousands of other couples use preimplantation genetic diagnosis, or PGD, they use a needle to physically remove one or two cells from the developing blastocyst and sequence its genomes. The result of this sequencing can tell you a lot about the developing baby. In theory it can tell you how healthy it is and a number of other traits it has. Practice is another story altogether.

To accurately read the DNA of a cell, you need lots of cells, hundreds or thousands of cells. Think *CSI: Miami*–you need enough blood to ID a killer. Same here: You need more than two cells. So to make up for not having much DNA, geneticists replicate the DNA they do have. This is called amplification. The problem comes when, like a game of telephone, small mistakes in the copies are preserved, then further compounded by later mistakes in the amplification. When multiplied millions of times, it becomes hard to trust that what the machine reads as a C is actually a C, and a G is a G, and so on.

While PGD catches big issues like chromosomal abnormalities, it's not so accurate at reading genetic code. Recent studies have shown up to 60 percent of single gene mutations could be missed by this technique. So a couple having to decide which embryo to implant based on PGD is using data that may not mean much to begin with. In other words, that embryo flagged with a rare disease may not have one, and the other way around. It's far from foolproof.

There are couples who have high-resolution genetic screening done on their embryos, but these are couples with a serious medical reason that justifies the risk of extracting up to 256 cells from the embryo. They sequence parents, grandparents, and other family members to statistically determine if a mutation found is actually there or an error. You can do this for a few genes but not all.

So the promise of easy PGD to select a designer embryo is suspect. Until we have a better sequencing or amplification technology, PGD will be limited to identifying gross abnormalities or highly specific traits known to the parents.

There are companies emerging right now that promise to predict the future intelligence and height of each candidate embryo by reading its genome. They're taking advantage of what people imagine is the future–investors included. But it can't actually be done accurately. We looked at creating just such a company, and the best fertility scientists in the world assured us it's not possible yet to do with any confidence. So companies shouldn't be making any such promises to expecting parents.

As I explain this to the women at the salon, they're surprised to hear it. They thought DNA sequencing was in fact foolproof. They thought that was the easy part.

One of the women in the salon asks, "But they're going to just go into the DNA of embryos and edit them, aren't they? Isn't that what the Chinese guy did?"

In 2018, headlines roared around the world that a Chinese scientist named He Jiankui used CRISPR Cas9 to edit two human embryos to become immune to HIV. Two edited little girls were born. Over the next months we learned more about the problems they faced. First, they used PGD to make sure the edits were accurate and where they needed to be. Of course, you now understand the problems of accuracy with PGD. Another issue with editing babies is called mosaicism.

A mosaic is a picture made of different pieces. Jiankui was trying to replicate a mutation in the CCR5 gene that occurs naturally in about 1 percent of people; it makes them immune to HIV. When Jiankui injected CRISPR into the embryo of an HIV-positive man and a woman, it scanned the genome for the CCR5 gene, which allows HIV to enter the cell. When CRISPR finds its target sequence, it stops and cuts the DNA in that spot, disabling the gene and closing the door to HIV. The reality is the cut doesn't happen immediately, and CRISPR activity is short-lived. Only about 80 percent of the cuts and necessary repair will happen on the first day, and then a bit more over the next two days. The result is that some of the cells will be edited and others may not be, resulting in a "mosaic" of genotypes. Jiankui reports that one of the twins born is a mosaic and still vulnerable to HIV.

Because of all this uncertainty, the next time someone tries the technique, they'll do it not on two fertilized eggs, but on a dozen or more. Then they would let only the most accurately edited eggs progress to an embryo. But even a dozen eggs likely wouldn't be enough to find one true success.

To enable a future of designer babies, a geneticist would need a hundred eggs, or a thousand.

To solve that problem, the final mind-bending technique being researched is called in vitro gametogenesis. If fully developed, the technique would enable millions of eggs per person, women to have children with other women, and even multiparent babies. That's even scarier than designer babies. Armies of super-soldier clones! Stealing the future babies of strangers or celebrities! Children with ten parents!

Don't worry. Not happening.

Remember how early embryonic cells turn into the different cells of our bodies? The timing of these changes is caused by just four proteins, called Yamanaka factors. These four magic proteins have the ability to turn cells into other cells. From stem cells into connective tissue or skin. Amazingly, they can also get cells to reverse direction, regressing from skin cells or blood cells back to stem cells. To use the tree analogy, we can transform a root cell into a trunk cell and into a leaf cell, and back again. This technique is called reprogramming.

In vitro gametogenesis is the technique of reprogramming a woman's or man's skin cells backward, not just back to a pluripotent stem cell, but to the Holy Grail of cells, an embryonic stem cell. And then reprogramming it forward, to go through the branches that lead to sex cells like egg and sperm. Voilà, millions of eggs from a cheek swab or blood draw.

Except this has been done only in mice. And doing this in humans is much more complex. So much more complex that it may be impossible as long as embryo studies are banned after fourteen days of division, because that is where we will discover the path to gametes, or gametogenesis. Because of this, many scientists are asking for the fourteen-day rule to be moved to twenty-eight days.

Kim Kardashian, the women in the nail salon, and every parent I know all want the same thing. A healthy baby and a career they love. Barring multiple scientific breakthroughs, Kim is pointing the way to have a baby later in life as safely as possible for the mother and child. A surrogate mother. Older parents have the means and stability to raise children and devote their time and energy to them. So they are turning to surrogacy.

I know people all over the world who are having children late in life and using surrogacy to help. One friend had two children five months apart with two different surrogate mothers. The mother was forty-two and found surrogacy to be a real solution to her biological clock. The $200,000 was no problem for this family. They would do it again.

An entrepreneur I met was born via surrogate because his parents are leading, high-powered doctors and his mother didn't want to slow the pace of her practice. He has a twin sister.

Surrogacy is a booming industry. Across Eastern Europe and Asia young women see surrogate motherhood as a way to make lots of money fast. So a white market and black market have sprung up, matching donors and mothers from around the world. Tears of joy from elated couples have been spilled as they receive their healthy new children. Tears of heartbreak, too, hit the ground as surrogate mothers are not treated well by their doctors or others in their society during the pregnancy. The donor couple and surrogate will never meet. Hopefully this industry can regulate itself to bring rules and fairness to all surrogate mothers. In America the rules are quite strict and vary state by state. In Ukraine, not so much. India and Nepal have banned commercial surrogacy altogether, leading to a thriving black market.

You can see this as a cruel shifting of motherhood from the rich to the poor. You could see this as a way for all women to get what they want, financial independence and the equality that comes with this power.

One of the women in the salon jumps in one more time with a question about when it makes sense to freeze eggs. Considerate of her age, I don't want anything I say to be interpreted as medical advice, so I put it in the form of a prediction.

"In the future, I think we'll see women freezing their eggs when they're relatively young. Men, too, will freeze their sperm."

The reason is that the older we get, the weaker our genetics. For men, every year older means an average of two more mutations are passed on to their children through their sperm. Women's eggs can somewhat repair sperm's DNA, but the older the egg, the weaker its ability to do that repair. Freezing gametes isn't necessarily a fix, because in the process, not everything goes perfectly, either. Cryopreservation can damage the cells any number of microscopic ways.

But this field of science is improving fast, and the odds of a frozen egg or sperm being healthy are going to rise rapidly. (The scientists get a lot of practice because it's common to freeze gametes before any cancer therapy.) Eventually, the benefits of using frozen-when-young will surpass the benefits of using fresh-but-old.

And if nothing else, freezing is insurance. It's not perfect insurance, but it's a shot.

This is the real future of fertility technology. Using surrogates and cryopreservation to shift time—all in the interest of healthy babies who grow up to have equality.

Saying good-bye to our new friends in the nail salon we walk out into the LA sunlight. We get in the car and drive off. My wife isn't convinced about surrogate motherhood for her. But we agree on one thing.

Designing babies isn't the future. Not just because we can't do it. It's because nobody wants it.

---

* To the idea of turning a skin cell into a human egg cell, you know how I wrote, "Don't worry, not happening"? And I was sure it wasn't possible? Well, after writing this chapter, I got to thinking, "Is there a different way?" And I started looking. And I found a scientist who had an idea. He's in our lab right now working on it.

28

Americans Now Spend Twice as Much on Health Care as They Did in the 1980s CNBC

Okay, I'm going to propose a very radical idea.

Everyone's been trying to solve our health care crisis. Maybe we don't need a brand-new plan. Maybe we can repurpose an existing plan that's already been written, like the Green New Deal.

And swap out green for blue, so it becomes the Blue New Deal. Blue like Blue Cross or Blue Shield.

I'm aware this sounds preposterous. So pretend for a moment that I'm not at all serious, and that this is just an interesting thought exercise. Let's explore the question of whether planetary health and human health have much in common. And let's evaluate whether the solutions to the former shed any new light on the solutions for the latter.

If you're not a fan of the Green New Deal, no problem. It works for the Paris Accords or Project Drawdown or whichever atmosphere-rescue plan you prefer. I'm just using the Green New Deal because it's got the catchiest name.

My hunch is that if we force ourselves to view health care reform through this lens, it's going to reveal new insights–and perhaps change our priorities.

I'm fully aware this is uncharted territory. It's an experiment. To win Super Bowl LIII, the New England Patriots didn't say, "Hey, let's borrow the game plan the Lakers used from the Showtime era!" And to beat back Al Qaeda, Gen. Stanley McChrystal didn't say, "Let's try Steve Jobs' plan to create the iPhone."

I know, it's outrageous. *Healing the planet can't possibly teach us anything about healing our bodies!*

I'm doing it anyway.

Think of it this way: Health care reform is normally boring. This won't be.

My hunch is based on this: Our planet, and our bodies, are a mixture of physics, chemistry, and biology, which rely on the same common atoms and the same scientific laws. Both get polluted. Both build up side effects from energy production. Both are dramatically impacted by microbes. Both require prevention and repair. Both are going to cost a lot. Both have been ignored for decades. Both are encountering problems we've never seen on this scale before.

So without going through every little detail of the Green New Deal, let's hit some highlights of this crazy analysis.

1. DECARBONIZE THE ENERGY SUPPLY

Oil is 87 percent carbon. Coal is 88 percent carbon. When we burn it, we fill the atmosphere with trouble.

At first glance, this has nothing to do with health care. But consider this: Of all the things we eat, the big troublemaker is sugar. Sugar leads to obesity, and in turn to diabetes, and in turn to NASH. Sugar is 37 percent carbon. It's the most carbon-rich thing we put in our bodies.

Just as Earth's thermostat has been dramatically jarred by the hockey stick of atmospheric CO_2, the human body's natural homeostasis has been overwhelmed by a hockey stick of C6 ingestion. Until a few hundred years ago, sugar was very expensive and rare; it was a fine spice. Then the industrialization of cane sugar, beet sugar, and high-fructose corn syrup paralleled our industrialization of fossil fuels. In 1700, the average person consumed less than four pounds of sugar per year. By 1900, that had risen twenty-five times. Today, in the United States, the average person eats 181 pounds of sugar a year. This may have leveled off in the last five years, but it's leveled off at ten times a sustainable level. Evolution did not design humans to eat so much sweet carbon.

If the first order of business in climate policy is to stop polluting our planet, then the first order of business in health care reform is to stop polluting our bodies.

In 1990, we had 6 million people with diabetes. Today, we have 29 million. That's a rise of 483 percent. You might think, "Oh, yeah, diabetes is gradually getting worse." No it's not. It's skyrocketing. It's a public disaster. We spend $200 billion as a nation treating diabetes.

Another 84 million people in the United States have prediabetes, a condition that if not treated leads to type-2 diabetes. Technically, 25 percent of them will have diabetes in five years. Seventy percent of them will get diabetes eventually. Which means we're sitting on a time bomb of trouble, financially. In just five years, we should have 50 million people with diabetes. It's going to make the opioid epidemic look puny. Prediabetes is like the methane stored in the permafrost. If we keep polluting our bodies with carbon, prediabetes will turn to diabetes ahead of schedule.

The real reason that health insurance premiums are skyrocketing is quite simple. The number of sick people is increasing.

Human movement has fundamentally shifted from one carbon source to another: We used to flex our muscles and burn calories to get from one place to another. We burned sugar. Today we sit behind the steering wheel. We burn no sugar, and lots of oil. The more we adopted motorized power, the more the planet suffered, and the more human health suffered. Just as we have to decarbonize our energy industry, we need to decarbonize our food industry.

We live at a time when there's very few regulations if you want to sell high-sugar foods or alcohol. But if you try to sell a medical remedy to those addictions, it takes ten years of clinical trials and hundreds of millions of dollars to wind through the regulations. It's ironic. There's a new sugary snack at the deli every month, no prescription needed. But the carb-blockers Precose and miglitol, as well as the alcohol blocker naltrexone, aren't on pharmacy shelves; they're stored behind the prescription window.

And the American government subsidizes sugar the same way it does fossil fuels. The average sugar farmer in the United States—there are 4,500 of them—gets $700,000 from the federal government annually. This costs taxpayers $4 billion every year.

In 2017, the *New England Journal of Medicine* warned that the "disease burden" due to poor diet causes more human deaths than any other risk factor. The United Nations declared that while one billion people are undernourished, more than two billion people are overnourished. Global obesity is following the lead of American obesity, growing at a pace that will overwhelm medical systems abroad—even in impoverished countries.

2. WE CAN'T AFFORD IT

That's what everyone says about the Green New Deal. It's also what they say about health care reform.

We wouldn't be arguing over climate change if dealing with it wasn't so expensive. If it was cheap to solve, everyone would be like, "Oh, yeah, let's fix it." The same is true for health care reform. Insurers don't want to be stuck with the costs. Hospitals don't want to be holding the bag. Doctors aren't taking a pay cut. Between them, we've got a Mexican standoff.

What we don't realize is: *We're already paying for it.*

In health care, everyone says more or less the same thing: "It can't go on like this. But I just don't see how it changes."

In the United States, we spend double the amount on health care per capita as other wealthy countries. Where is all the money going? It's not that we visit the doctor more often. Far from it. In Japan, the average person visits the doctor three times more often than us. And they're spending only half what we spend on medical care. Even in Canada, people see the doctor twice as often. Nor is it the case that we stay longer in hospitals. We stay less.

But our prices are just so much more expensive. Surgical procedures like knee replacements and C-sections and appendectomies are all double the cost of elsewhere; so are MRI exams and colonoscopies. Drugs cost about double, too. Xarelto, the blood thinner, is twice as high. So is Humira, for rheumatoid arthritis. The common chemotherapy drug Avastin costs nine times more in the U.S. than in the UK.

The harm this is doing to our economy is not obvious. It's a whopping side effect, but a side effect nonetheless. Companies are paying more than ever to insure their workers, so there's nothing left to offer as wage increases. It's drastically suppressing real wages and salaries, especially in the lower and middle classes. In one analysis of Federal Reserve data by the Mayfield Fund, it showed that the cost of hiring an employee has steadily risen–but the wages paid to the employee (factoring out inflation) did not rise. The difference has largely gone to the health system. This has been going on for four decades.

What it adds up to is stunning: Worker wages could have been 60 percent higher today, if we had kept health care costs in check.

Another way to say it is: American workers will never see another pay raise if health care costs keep siphoning off the money.

3. YOU CAN'T CHANGE WHAT YOU CAN'T MEASURE

The Paris Accords require all countries to accurately estimate their emissions. They have two standards, one for the developed world (which is stricter), and one for the developing world (which allows higher emissions). We can't model the atmosphere in supercomputers without accurate data. There's a lot of confusion about whether cows or rice or permafrost is the main source of methane. Deforestation and reforestation are happening at the same time. Natural gas is pretty clean energy, except if there are leaks—and there are a lot of leaks.

We're hopeless unless we can measure it. We can't police it unless we can detect it.

And I'm going to say the same goes for our health. In health care, it's *all* about catching it early. You catch cancer early, you'll be fine. You catch it late, and you're dead. Dead *and* broke.

This goes for disease after disorder. Cardiovascular disease. Stroke. Kidney disease. Depression. Parkinson's. You name it. The earlier we detect autism, even, the better the outcome.

Early detection is a twofer. You recover from the disease and avoid catastrophic health care costs.

To pull this off, we need to stop playing the telephone game. The telephone game is where one part of the body talks to another part of the body, who repeats it to a third part of the body, and so on. And at the end of the line is the doctor, reading the report. What she gets is a very fuzzy, out-of-focus version of what's really going on.

The vital signs that doctors test are excellent at telling the doctor if you're about to die. They're *vital* signs. Life signs. But they're not so good at telling the doctor if you're going to have a rheumatoid arthritis flare-up in three days, or if the genes in your knees slowed down expressing some critical protein. Small cancer tumors are almost impossible to detect. We walk around with diseases, sometimes for years, before our symptoms materialize.

The goal here is to tap the telephone. And to listen in to the way cells actually talk to each other. Right now, these signaling molecules are not testable in regular lab tests. Cells send messages to each other with exosomes, small RNA, antibodies, specific immune cells, analytes, and hundreds of itty-bitty, small molecules. It's a kind of code that not even the NSA can snoop on. Most commercial blood tests can barely read any.

MIKEY THE CELL: Hey man, I don't feel so well.
MIKEY'S TWIN BROTHER: Don't cough on me! You're going to get me sick.
MIKEY: Uh-oh, here comes some immune cell.

TWIN BROTHER: Don't worry, he can't see us.
IMMUNE CELL: Stop or I'll tase you with cytokines.
Zap zap zap zap zap
TWIN BROTHER: You killed Mikey!

Now in theory you might have a fancy health care service that brings you in every year and runs a lot of expensive tests. It's better, but it's not the answer. I can guarantee you they are not spying on the actual cell-to-cell communication. What they're getting is at best thirdhand telephone.

Let me give you an example. And let's stick with diabetes, since that's the time bomb. To diagnose diabetes, most doctors use a test called "fasting glucose." It's not very accurate. If you've got fancy health care, your physician would use a more expensive test for "glycated hemoglobin." It's a good test–it has some accuracy issues in edge cases. But if you want to detect diabetes *before it happens*–which is the goal here–you need these:

Fructosamine, Glycated albumin, 1,5-Anhydroglucitol, Adiponectin, Fetuin-A, alpha-ketobutyrate, L-alpha glycerylphosphorylcholine, Lipoprotein(a), Triglycerides, HDL-C, Ceramide, Ferritin, Transferrin, MBL-associated serine proteases, Thrombospondin 1, GPLD1, Acylcarnitine, miRNAs, CRP, IL-6, WBCs, Fibrinogen.

I put the list in small print because it's the sort of thing you shouldn't actually read. Unless you're a doctor. But doctors cannot detect diabetes before it happens, because they can't measure all those biomarkers. You can get at those details only in the world's best research labs, and it would cost thousands of dollars to run all the varied tests.

This field of high-resolution biomarker detection is called multiomics. It's just starting to show some incredible promise. For example, to detect nonsmall cell lung cancer, in the very early stage, physicians were measuring a molecule called pro-surfactant protein B. But then they found they could improve accuracy dramatically by also looking at the metabolite N1, N12-diacetylspermine.

I'm hitting these examples pretty hard, and with detail, to make my original point more convincing: Vital signs can't do it. Basic blood workups can't do it. Artificial intelligence can't do it. To fix health care, we need to be able to measure trace compounds that we've never measured before.

Diagnosing patients is hard. Symptoms are just the first clue. For instance, let's say a doctor knows a patient has something wrong with her pancreas. But is it pancreatic cancer, or just chronic pancreatitis? To make this diagnosis, physicians rely on a test to look for a cancer antigen called CA19-9. To do it earlier, with accuracy, you need CA19-9 *and* these nine other analytes:

> Proline, Sphingomyelin s17:1, Phosphatidylcholine, Isocitrate, Sphinganine-1-phosphate, Histidine, Pyruvate, Ceramide, and Sphingomyelin d18:2.

It would take thousands of dollars to run the multiple tests to capture all that data. Most tests just capture a dozen or so analytes. We created a company, with a scientist out of UCLA, that found a unique way to hack testing equipment–using a combination of engineering and physics and computation–to get thousands of biomarkers out of a single low-cost test.

It opens the door to the possibility of a new era of medicine. For a low annual cost, physicians could diagnose disorders before symptoms appear. This would drastically reduce health care costs, and vastly improve the health of everyone.

We created several companies that can spy on cell-to-cell communications. They can read small RNAs, they can stratify and count the 4 billion immune cells in your body, and they can detect autism in a newborn baby. They're part of a fleet of many new companies that are finally going to revolutionize health care.

My biggest concern is that the politicians and the policy wonks who stitch together our health care policy don't realize we are on the verge of this breakthrough. I'm afraid their reforms will lock in the old way–just before we've finally cracked the code.

4. LAWYERS

Most people don't know that a couple years ago, the oil industry came out and suggested a big new tax on their industry to address climate change. It was led by Exxon and Shell. They proposed a carbon tax of $50 per ton of CO_2 to deal with climate change. To make sure everyone knew they were serious, they got endorsements from two former secretaries of state, James Baker and George Shultz, as well as former Treasury secretary Henry Paulson.

Yes–this is a *real* story. They wanted to tax themselves.

We imagine the oil industry is fighting climate regulation, but in fact they've proposed the simplest and most concrete plan of them all.

So why didn't it happen? Because they wanted–in exchange for the carbon tax–to be absolved of any financial liability for decades of polluting the atmosphere. They didn't want to be sued.

In my opinion, that deal should have been struck. But it went nowhere. No way were the lawyers going to be prevented from suing. Quietly, the plan was scuttled.

EXXON: We'll pay for solving climate change.
SHELL: Just don't sue us.
LAWYERS: Nope.

When this happened, climate got a little taste of what health care has been dealing with for fifty years. Lawsuits cost a lot of money. The risk of lawsuits costs far more. Every hospital has to be run like an ultrasafe, risk-avoidant facility. Every procedure has extra steps and extra documentation. Far more tests are ordered than are necessary, just in case. Medical centers and hospital systems are not racking up all these costs to get rich. They're not evil. They do it to avoid being sued.

The United States is the only country in the Top 50 on the Human Development Index that doesn't have socialized or universal medicine. So it's tempting to think, "We should do exactly what everyone else does." Except none of those countries has the kind of litigious culture we have. If the government paid for health care, then everyone would be suing the government. If our government actually ran our health systems, we would have lawsuits out of our ears.

Legal reform is critical to make *any* health care reform work.

5. THE GERMS WILL SAVE US!

Gradually we're waking up to the idea that microbes are going to be our best friends in saving the planet. We used to say that forests are the lungs of the planet, and that they capture all the CO_2. Now we know that it all depends on which microbes are part of that forest soil's phytobiome. One of the big players is a fungus called ectomycorrhiza, which radically increases CO_2 sequestration in deep soil. Ectomycorrhiza is doing over twice as much as the trees.

Remember how Arvind went to Iceland to study their carbon sequestration method? Well, ectomycorrhiza does that naturally, for free. All we need to do is grow it and spread it in the soil. Oh, and stop overfertilizing. The nitrogen and nitrous oxide released by fertilizer kills off ectomycorrhiza. We've been silently killing off the real lungs of the planet. Those lungs have been smoking nitrogen.

There are microbes that suck in CO_2 and spit out oxygen, and all we need are sulfur vents. (We have a ton of those.) One of our companies sucks in CO_2 from cement factories and turns it into fishmeal for aquaculture. It was microbes that rescued the Gulf of Mexico after Deepwater Horizon exploded. Microbes can turn *anything* into *anything*.

Let me get back on track here. I could write a whole book about the microbes that will save us from climate change. But this chapter is about human health.

Nevertheless, the same principle applies.

Did you know that the bacteria on your skin are critical to preventing heart disease? Did you know that bacteria in your mouth migrate up to your brain and contribute to Alzheimer's disease? And that a virus we pick up as children goes dormant and hides in our brains, where it is reactivated by stress and causes many mental health disorders?

For about a decade, we've been hearing that the microbiome is important. That's like saying lungs are important. Our microbiome weighs more than three pounds, and it's the invisible vital organ.

Almost every biological pathway relies on a chain reaction of conversions that begins with metabolites produced by the microbiome. The microbiome is our natural drug factory. To fight inflammation, we need the histamines it makes. To regulate gene expression, we need its short-chain fatty acids. To make energy, we need its chylomicrons. To regulate blood pressure, we need its tyramine. To fight DNA damage, we need the microbiome to make alpha-tocopherol.

We rely on the microbiome more than we rely on Amazon or our iPhones. To lose weight, we need its deoxycholate. To prevent atherosclerosis, we need it to produce zonulin. To protect our brains, it cranks out glucagon peptide 2. I could go on and on.

Not only does the microbiome crank out drugs on a daily basis, it explains why a lot of the drugs given to us by physicians work on some people but not others. You know how, when you visit the doctor, you have to fill out those forms asking what other medicines you're taking? That's because drugs interact with each other. Except there's no form for your microbiome to fill out, to report what drugs it's been filling our bodies with. (Though with multiomics we're getting there—see point 3 earlier in this chapter.)

We understand the microbiome about as well as we understand the weather. We understand a lot of the pieces, but we can't prevent hurricanes or heat waves. The microbiome's 100 trillion cells have a lot of tricks. They mutate a lot. They rely on each other. They can also swap DNA. So we have to be very careful about false confidence in thinking we can just mess with it.

However, what we have learned in research is now just starting to inform what goes on at your local hospital. The more we grasp the interactions, the better health care is going to get.

6. INCENTIVIZE THE FUTURE, DON'T REGULATE THE PAST

The Green New Deal calls for a trillion dollars of innovation. It recognizes that our old technology is causing the problem. So we need new technology.

My big beef with health care policy is that it's trying to fix the system we have today, rather than create a vision for the future of medicine–and help that get here faster.

The Affordable Care Act of 2010 at least did that. It offered hospitals $30 billion in incentives to adopt electronic patient records, over multiple phases. In this one dimension, it dragged hospitals into the modern age. (Full disclosure: I've worked on electronic medical record reform with several CEOs of both hospital systems and software systems.)

Health care is a deeply entrenched industry. No matter how we, as a society, decide the question of "Who pays for it?" the *real* priority ought to be incentivizing future technology to get adopted.

I'm not sure what it says about the state of health care policy that this crazy experiment, the Blue New Deal, works just about as well as any other policy plan at reconfiguring health care. It worked so well that I'm thinking of trying it some other places. Campaign finance reform? You never know. Tax reform? Sure.

Or maybe the lesson is that health is health, and biology is biology. The planet is alive, and so are we. Two degrees Celsius of global warming doesn't sound *that* bad. Until you think what that might mean if your body temperature was raised by two degrees Celsius. You'd be permanently running a temperature of 102.2 degrees Fahrenheit. Symptoms include persistent vomiting, seizures, and hallucinations.

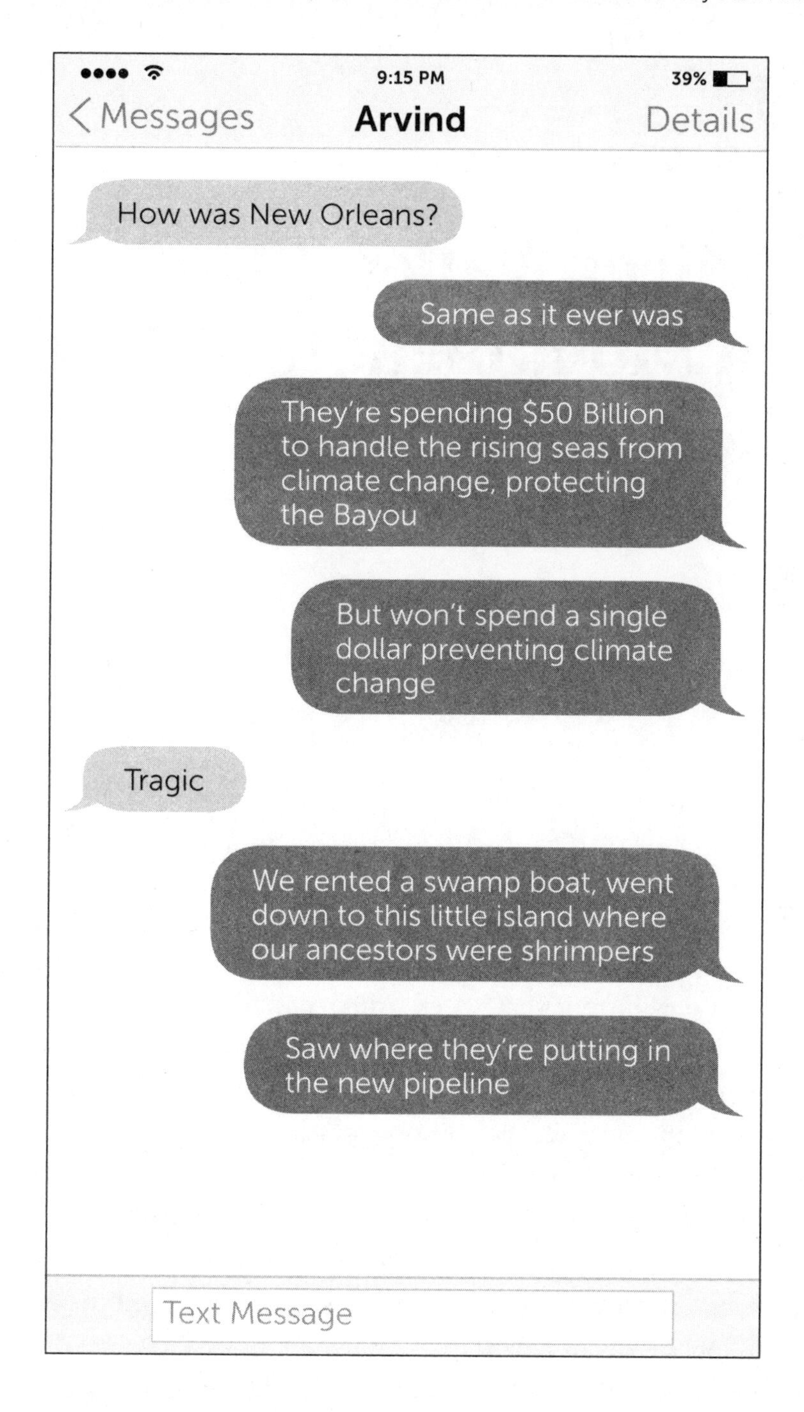
9:15 PM
39%
Messages
Arvind
Details
How was New Orleans?
Same as it ever was
They're spending $50 Billion to handle the rising seas from climate change, protecting the Bayou
But won't spend a single dollar preventing climate change
Tragic
We rented a swamp boat, went down to this little island where our ancestors were shrimpers
Saw where they're putting in the new pipeline
Text Message

29

In Silicon Valley, Plans for a Monument to Silicon Valley New York Times

Silicon Valley has no history.

Now, this isn't technically true, but it's psychologically descriptive. A handful of Valley histories have been written and filmed; I've been a part of several. So it's not that they don't exist. Rather, it's that they are not looked to for guidance. Another way to say it is that our history is ignored. Not willfully ignored–it's more like our history goes into a black hole of cognitive oblivion. We don't look backward for answers. We don't replay the tape.

In Silicon Valley, when someone's career goes off track, or a venture firm fails to return its fund, nobody examines the past for answers. Nobody dusts off that hardcover history of the post-dot-com period to search for answers. Nobody sits in a dark room, watching black-and-white documentaries about Nolan Bushnell and Xerox PARC, listening for inspiration to solve the problems of today.

All over the rest of the world, history is important. Memorials are built. Grudges are remembered and retold. Sins are immortalized. Battlefields are made into parks. Sacrifices are commemorated in carved stone.

Silicon Valley has never bothered to put up a plaque. San Jose has threatened to erect a statue in a park, but most people think it's wrong-minded. *We don't want a history.*

No ground is hallowed. If you visit a startup, nobody there cares about the startups that used to occupy the same space a decade prior. Blank stares.

This rejection of history's conventional role creates a distinct psychology. Nothing is sacred. Nothing needs to be preserved. Nothing is off-limits. If we are doomed to repeat our mistakes, so be it.

If Silicon Valley went to a shrink and rambled on the couch for a while, we would be like one of those Oliver Sacks oddballs. The psychiatrist's diagnosis would read: "Patient S.V. suffers from a syndrome marked by time dysphoria and antithymesia. S.V. is uncomfortable with the present and expresses desire to live in the near to distant future. Reality is considered temporary, almost an illusion. Patient holds beliefs that all will change and be replaced."

So, if history is no guide, then what is? What serves as its *conscience*? Where and what is its subconscious process?

I'll restate the question because I mean it in a particularly philosophical way. According to Carl Jung, the collective unconscious is our inherited history, stripped clean of details and forged into archetypes. Even if we forget the details of history, we absorb a blurry gestalt of history through archetypes. These archetypes act, psychologically, like a kind of magnetic north. We can cling to them or we can rebel against them, but either way they provide orientation.

Every industry has its own idiosyncratic ways of building trust and getting business done. Trust is the backbone of all dealmaking, worldwide. Silicon Valley is no different. Don't listen to the myths about meritocracy. In the *real* Silicon Valley, trust is built at group houses, at poly communities, at the Russian baths where they whip you with leafy twigs of juniper and eucalyptus, or while kite surfing (in Maui). To deepen these connections, there are no-pants parties, and self-care meditations, and a floating-on-the-water version of Burning Man with renewed idealism. People have teahouses in their basement, where you might run into a VC you know–in a mermaid costume. The word "transformation" is used a lot. People take baths–with sound.

Every night, somewhere in Silicon Valley, people are gathering. People go and spend a few hours, and when they walk out, they often feel unsure what it was all about. They often feel like they barely had one insightful conversation all night. Maybe none. Into normal conversations, people drop half sentences and snippets that only half make sense, hoping someone sees insight in them.

"Will the metaverse be centralized or decentralized is the ultimate question."

"Our DNA can be sent to deep space hundreds of years before we can get there."

"We won't solve anything until the social status system values internal states of mind as much as it values external possessions."

On a good night, it's a kind of intellectual primordial soup. On other nights, it's emptying the subconscious garbage can of that week's stresses. The difference between them is chance. You never know, when you show up at the door, what kind of night you'll get.

It was on just such a gamble that we found ourselves attending a party one night, hosted by His Royal Highness Prince Constantijn Christof Frederik Aschwin of the Netherlands. We didn't really *want* to go. But we felt obligated. The prince has been a big promoter of the startup culture in Amsterdam. It's ironic in a way that the Netherlands probably does understand.

The soiree was at an old stucco manor on Alamo Square, called the Archbishop's Mansion. The main floor has all these rooms. Wealthy travelers stay in the rooms upstairs. There was a line to get in. When we got to the door, they gave us a map, and said, "You are entering the year 2029."

Arvind, you should tell this part.

Happy to.

The whole shindig was interactive performance art. Trying to awaken our future consciousness. It was striving to be time-trippy, and it couldn't really deliver on that, but a real effort had been put into it. It wasn't lame. Through these French doors, the first room was the death room. We just hung out there. A woman was performing this monologue about how you die and get reincarnated as other people, and every time you die you become a new person, until finally you've been everybody, and then you're free because you have complete empathy.

Then we went into the feminine room. This actor playing a Nigerian princess welcomed us, served us soup, and talked about how disconnected we've become. And then there was the ocean room, which was littered with plastic everywhere, and a woman in a robe was wearing a plastic buckyball molecule around her head and reciting climate change statistics. It was Burning Man lite, without the dust bowl. I wandered into yet another room. The prince was there; he wasn't attached to a flock of aides and handlers. He said hi to people all by himself.

The essential oxymoron of all Social Purpose Parties is the tension between the Party and the Cause, and how that translates into behavior and affect. It's often unclear: Are we supposed to enjoy ourselves and get silly, or are we supposed to be somber, earnest, and intellectual? Social Purpose Parties often get stuck in the uncanny valley between the two, which dooms them to be forgettable. The prince's gala managed to avoid this trap; it pretended to have no particular social purpose—to just be an experience, ten years in the future.

We'd been there an hour and I'd met the prince, but it was still hard to tell if this was a good night or a bad night. I hadn't yet had an insightful conversation, and I was close to giving up it would happen. I appreciated the way we could just swim room to room and float in this Halloween-like ode to 2029. It wasn't force-fed to us. But I could only take so much of it—I was considering skipping out. Then I met this physician, and I talked to her a little bit about what we do at IndieBio, and she ran away, saying I had to meet Tom Chi.

Tom Chi was one of the founders of Google X.

The physician comes back with Tom Chi. My memory bank would not have even stored the evening if it wasn't for meeting Tom. It was all this typical Silicon Valley craziness, but if we hadn't gone—and hadn't humored it—we never would have met Tom.

This was the moment it suddenly became a good night.

That's right.

You want to explain climate change, the Tom Chi way?

You do it.

Everyone's got their way of explaining climate change. And the way you explain it always shades it a bit, steers you to certain solutions over others. That's true for Tom, too—but in his case, the way he shades it helps us really understand both the size of the challenge and realistic approaches to solving it.

Tom looks at solving climate change as a mass transfer problem. We have all this mass in the atmosphere (carbon) and we need to get it back into the earth.

Just how much do we have up there? Think about the size of one human. Just going about our lives, each human averages a little over a ton of carbon going into the atmosphere, *every year.* That's like blowing up a car, exploding it, and vaporizing it. All 7.5 billion people on the planet doing it, every year. Nobody is carbon negative. Nobody is carbon neutral. We're *all* part of the problem.

So even if we magically stopped doing that, there'd still be a trillion tons too much carbon in the atmosphere.

The ocean has absorbed 582 billion tons of carbon dioxide. The ocean has also absorbed 93 percent of the excess heat. If we didn't have oceans to absorb heat, the planet would be fifty degrees hotter, Fahrenheit. This absorption changes water acidity, and the organisms that grow in the ocean are really sensitive to this change in acidity. If we continue, all corals will be extinct in about thirty-five years. No shellfish will be able to grow in the ocean in a hundred years. The last time this happened was 56 million years ago, and it caused mass extinction of marine life.

So it *can* sound pretty hopeless, but that's not how Tom sees it. Rather, this just tells him the precise magnitude of the problem. Tom laughs off a lot of "sustainable" companies as barely making any real dent in the problem. We need solutions that work on this titanic magnitude.

Planting forests is one option. Right now we cut 6 billion trees a year. When a typical tree is at maturity, it's got over a ton of carbon in it that it sucks out of the air. We'd need to plant one trillion trees, and as they grew, they'd pull the carbon out. That's a lot of trees, so Tom built a robot drone that can plant 120 trees a minute. His goal is to plant 20 billion trees a year. He can do it for about $80 million a year; it'd take nine thousand drones and a staff of about 450 people.

A solution that's even easier is to recarbonize our farmland soil. Farm soil today has only about 1 or 2 percent carbon in it, because we've been burning off our topsoil for centuries with farming, releasing all the carbon. Farms that practice no-till regenerative agriculture can increase the carbon to more than 8 percent. Healthy soil is dark because of the carbon. There's a massive amount of biomass in the ground–microbial life, nematodes, critters.

Here's the thing: We already have 9.9 trillion tons of carbon in our soil right now. We just have to increase that by 11 percent, and we'd pull a trillion tons of carbon out of the air.

So a trillion tons of carbon is both a really big number but also a really doable number if we attack it intelligently.

Tom lays this all out for us. It's the most incisive, penetrating, no-bullshit calculation of climate change I'd ever heard. And all because of the prince and his insane party.

Meeting Tom gave us a real lift of spirit.

Having Tom as a compatriot created a kind of second renaissance at IndieBio, where we had the courage to invest in solutions that were going to rely on carbon credits for income.

Tell them about olivine cakes.

Sure.

Ever since I got back from Iceland, it's haunted me that we can speed up the conversion of captured CO_2 to rock with the plentiful mineral olivine. The entire west coast of North America has olivine under the crust. Gavin Newsom, our governor, had just approved new carbon sequestration credits that were the highest in the world–\$150 per ton of CO_2.

In Iceland, they took the natural fifty-year conversion to stone and sped it up into a two-year process. At IndieBio, we wondered if we could speed it up even more. We had this naive, somewhat fantastical idea–if we could get the conversion down to a couple days–we could just make bricks, or "olivine cakes," to sequester CO_2. We wouldn't need underground storage. They could be used in buildings, or even piled up as a monument. Or dropped in the ocean to gradually weather and become the shells of oysters and clams.

In Massachusetts, we found a team of engineers with a similar vision. They'd been struggling to get any investor interested. We agreed on a design for a bioreactor using pumps, lasers, and LED lights that would speed up the conversion so much, we could (according to the math) make an olivine brick *in just two hours*. We funded the engineers and brought them to California.

The rational investor never would have wasted an evening at the prince's party. But the subconscious process of Silicon Valley–the *real* Silicon Valley–had worked its irrational magic again.

1:01 AM
93%
Messages
Arvind
Details
Signing Friday
Oh man. Leaving your baby.
Setting it free.
But it is hard, ain't gonna lie.
We'll be fine.
We showed the world a new path. And thousands of scientists are starting the walk. I need to make sure they don't die of starvation.
Sand Hill Road is going to be so tame compared to Jessie Street
Sacrifice worth making
Text Message

30

Forget Terrorism, Climate Change, and Pandemics: Artificial Intelligence Is the Biggest Threat to Humanity Newsweek

I remember holding a human brain in my hands when I was fourteen or fifteen. It was in the autopsy room of UCLA's medical school, where I hung out over the summer at my dad's lab. I'd wander the darkened mile-long halls late at night, braving strange noises and vibrations coming from inside locked doors. I could feel the zombies coming to life, I could see them in my mind. So one night when I went down my usual route and took an unusual left turn, I found a door wide open. I snuck up to it. And stuck just enough of my head out to see in with one eye. I saw waxen gray bodies laid out on metal tables, dead. They had their torsos sliced open. Their arms were missing patches of skin with their muscles and veins clear to see. It was nothing like an anatomy book. The body parts were not various multicolor hues for easy identification. They were shades of gray, dull yellow and white and greasy. Like steak that had been forgotten in the back of the freezer.

"Who's the kid?" a man said, breaking the dull drone of the HVAC system. Horrified, I turned to run, but he stepped out and grabbed my arm. "What're you doing here?"

I stammered. "Uhhhhh."

I explained my dad was a professor and I was bored walking around. Turns out he was a medical student. And I had found the anatomy class. I was fascinated by anatomy at the time. I had dissected a rabbit and mouse in school already. So I knew the faint formaldehyde smell as I walked into the bright room, following the med student. His team were huddled around a gaunt old man lying on the table, his resected brain sitting beside his open skull. I asked questions, they answered. I put on gloves. A real Doogie Howser for fifteen minutes. They let me hold it. Waxy. Pale. Off-white. It blew my mind. I believed I was looking at the soul of this elderly gentleman. It hit me hard. All his triumphs, fears, failures, loves, losses, tears, and joys in two hands. I gently put it down, thanked the students, removed my gloves, left the room, ran down the darkened hall, and sobbed the whole way back to Dad's lab.

Later that night at home, I flicked on the power of the computer my dad had purchased from a guy in a garage. We built it from parts, so I got to see what was in it and how it went together. A motherboard with a CPU and RAM chips. Floppy disk drive. Monitor and keyboard. No mouse. He used it for making graphs. I used it to play text games, the kind that said, "You are in a dark room. Your sword begins to glow. Do you a) leave the room b) raise your sword." Blinking green cursor.

The personal computer began a revolution I lived through, though it didn't feel like it at the time. I suppose all revolutions are muted to their participants.

My dad and I would add RAM every once in a while. They were little chips with silver legs, like little rectangular beetles. The legs fit into slots on the motherboard of the PC. Random-access memory was the computer's short-term memory. This is like the memory we use when trying to remember a phone number before dialing it. I always have to repeat a phone number out loud over and over until dialing or I get it wrong for sure. And then I forget that number forever. Our PC clone in the '80s had 256K of memory, which meant it could hold more than 65 million phone numbers in its head at once. It sent these over to the CPU to calculate, and the results were translated to the monitor, glowing a bright green. While it was doing so, a red LED light would glow on the front surface for a split second or sometimes much longer.

When that light was on, my sister and I would say that the computer was "thinking." Thirty years later, we still do. The connection between computers that think for themselves and computer parts that mimicked our brain was lost on me at the time. For computer scientists, "thinking computers" isn't just a metaphor. It's a life goal.

The 1950s was the heyday of physical biology. Brain structures were being dissected and analyzed part by part. Neurons went under the microscope. Biologists saw that neurons interacted with each other, and the more they did so, the stronger the connection became. This idea became known as the neuron doctrine, which is the basis of neuroscience today.

Psychologist Frank Rosenblatt had the idea that mimicking the neuron doctrine in a computer might get it to learn the same way humans do. So he built a computer in a completely different way than our family's IBM PC. Instead of a CPU or RAM he created two layers of neurons out of circuits he called Perceptrons. Instead of a keyboard he used a camera. The idea was to show shapes to the camera and see if the computer could learn to classify them as triangle, square, and so on. Key to this approach was the Perceptron.

The Perceptron was a simple circuit that sent a signal only if the two signals coming into it reached a threshold level. The strength of the signals from the connections could vary. This difference in signal strength was the heart of the system. It allowed the network of circuits to adapt and change based on what it was told was right. So Rosenblatt then showed the camera a triangle, and input that it was a triangle. He kept doing this until the connections between the camera and the Perceptrons correctly identified triangles. He then set up the test. He showed the Perceptron machine a card with a circle, then a triangle. The machine's red light glowed.

It worked.

On July 8, 1958, the *New York Times* ran an article saying that later versions of the Perceptron would be able to "recognize people and call out their names and instantly translate speech in one language to speech or writing in another language." It went on to say "it would be possible to build brains that could reproduce themselves on an assembly line and which would be conscious of their existence."

Seventy years later we are still waiting for self-aware computers. But the rest they got right. The idea of the Perceptron evolved into modern machine learning by using more layers of neurons and new ways of training the network. Feeding the data forward, then backward again. Taking the results of one neural net and feeding it to another. There is now a large variety of algorithms, but they all use the concept of Perceptrons at their heart.

Cutting-edge research in artificial intelligence today is about getting computers to create new goals for themselves and seek ways to attain them. This approach has led to the development of deep-learning techniques that have created programs that defeated grand masters at Go and chess, and fighter pilots in aerial dogfights. Today's computers have billions of transistors to process the data they ingest. They are trained with a goal in mind, on millions of games or dogfights, and they exhibit nonhuman creativity, trying things humans never have.

But they can't learn in the way a four-year-old girl or boy can. A boy or girl can learn from just a few examples of "car" or "dog" and easily generalize far beyond those examples to understand all cars and dogs. It's been estimated that by the age of six, kids have ten thousand to thirty thousand object categories in their mind, despite having only seen a handful of examples in their life. They can draw new and original hybrids between objects. The ultimate test for computers–along these lines–would be "one-shot learning," where a computer can intuit what something "is" (a triangle, a Ferrari, a hot dog stand) from a single experience with it. But computers today are a million shots away. What they do is, yes, surprising–but it takes millions of examples to train on before it can determine that a triangle is a triangle, and a car is a car, and a pedestrian carrying four bags of groceries in the middle of the block is–yes–a pedestrian, despite the odd profile and location.

This is a fundamental problem for artificial intelligence to solve if it hopes to achieve even a four-year-old's intelligence, or even common sense. Common sense is easy for us but hard for computers. Common sense is the ability to apply learned principles to a new situation accurately.

There's an important distinction here between two terms of art that sound phonetically similar: interpolation and extrapolation. To interpolate is to bridge the gap between two categories. A computer can interpolate what an animal might look like, for instance, if a rhinoceros bred with a horse. Just merge the two, conserve the notable features...red light...red light...maybe you get a unicorn-type beast, with armadillo legs. A nickel is what you get when you interpolate between a dime and a penny. Computers can *do* interpolation.

But extrapolation is where computers get stuck. Extrapolation is common sense. It's the ability to make decisions and act reasonably in an entirely new situation never before contemplated, nor seen.

"What do you think?" I ask Peter Schwarzenbauer, board member of BMW Group. We are standing together in the basement of IndieBio's lab and we are staring at a bunch of spikes running across a monitor. The spikes are coming from electrodes that are wired to a series of human neurons grown in cages and interlinked with each other and being warmed by an incubator at 37°C. The spikes are electrical signals from the neurons when they fire. They are happening fast, ten to twenty a second and at random. We are looking at a primitive artificial brain with sixteen neurons, same number as a dragonfly, hooked up to a computer.

Peter turns to me with raised eyebrows. "Not sure this will work. But a fully autonomous vehicle's AI systems still cannot handle all the edge cases a driver would experience," he tells me. "It's hard to get all the data to tell the computer how to react in any specific case, like a foggy road in the rain or another driver making erratic moves."

I reply with a half question half statement. "You mean an AI car needs common sense?"

We both laugh. "Yes," he says. "When to pull over in bad weather and when not to. It is truly common because we all do it, maybe not perfectly, but driving is common to nearly every human."

We look back at the screen with pulsing spikes.

"Even though this brain is only as big as a dragonfly's, maybe it could process and store these edge-case scenarios for the driver," Peter says. "But this system is way too early for us to consider." He smiles. I can't blame him. This is uncharted territory.

If the company wired a billion neurons together, connected this network to cameras and a computer trained to drive, and let it run, would it learn common sense? I don't know. But I know who to talk to. He is famous for building a brain interface company and revolutionizing aerospace. He lives three hundred miles south, in Malibu. And he loves cars. Tony Stark.

So I rented a Tesla Model S and drove south. Once on the I-580 I flicked on the autopilot. The cameras and AI brain of the car took over. It was a beautiful fall day. In the Bay this meant eighty-degree temps and blue skies. There are cameras, radar, and ultrasonic sensors all over the car. These are its eyes. I set the speed at 75 mph and took my hands off the wheel. It's my first time. It's weird. The car's eyes are ingesting pixels and radar signals thousands of times a second to reconstruct a 3D model of the world around it. The AI picks out the cars around us, the trees, the lanes, and keeps us straight. It's like Rosenblatt's Perceptron Mark I but with billions of circuits instead of five.

For a while, I look around for edge cases. Situations, even objects, that the Tesla computers might never have seen before. But on I-5 between SF and LA, there's rarely anything eventful.

But the car still did not have common sense. It didn't guess other drivers' intentions as we do countless times a day. So it got stuck trying to make a lane change or didn't allow another car to make a lane change. It flashed a blinker, then stopped. Granted, humans do, too. But at least they know they are being assholes.

North of Bakersfield, I am bored enough to let my mind roam. I find myself wondering what the vehicle would do if it suddenly started raining frogs. That happened in Calgary last year. Biblical. Book of Exodus. I guess they get sucked up into the sky by a tornado or something. A true edge case. Common sense says pull over. Pull over and pray. Not sure what an AI would think. But I'm pretty sure the red light would be on for a while.

For decades, we imagined that AI would be amazing. Now that it's here, most people don't like what it's wrought. AI has been more raining frogs than parting waters. It's been seven years since AI could first recognize cats on YouTube. We are hard-pressed to name more than a few truly "good" examples of artificial intelligence. In India, an AI compared all the photos of kids in orphanages against all the photos of missing kids. It was able to find 2,930 matches and reunite them with their families. But on the backs of this feel-good story, India is launching facial recognition across the country. Even "good" AI accomplishments lead, ultimately, to surveillance states.

Most of my friends have come to fear an AI future. They entertain wild ideas like AI taking over, not because those ideas are reasonable, but because the world around us seems to have fallen into chaos. It *feels* like AI is making a mess of it. Plugged into social networks, AI seems to have obliterated the middle and pushed everyone to the poles. We are clickbait targets. Sensationalism rules the day. Racism, hate, and scapegoating are the new lingua franca.

But my gut tells me AI just isn't that competent and won't be for a long time.

I'm not wowed by AI's powers. Even as it takes me over the Grapevine and down into the San Fernando Valley, where I take the wheel again, for memory's sake. This is my home turf.

I arrive in Malibu. Driving on Highway 1, I take in the glittering Pacific Ocean. I used to skip class from Van Nuys High and go to this very beach to skimboard and surf. I pass the McDonald's we used to eat at afterward. Now I am pulling up to Tony Stark's clifftop mansion. The gate has no code. It just opens as I pull up. I drive in. A man in a black suit with dark sunglasses and an earpiece greets me at the end of the driveway. A rounded, low-slung house of white concrete and glass set into the cliffside spreads out in front. I can see through the transparent twenty-foot front doors into the foyer. I get out of the Tesla and the man in black takes it away. I open the door as instructed, and a tall blond woman meets me. "Ms. Potts. Nice to meet you, Arvind. Thanks for coming," she says.

"Thanks for having me."

"Ahhhhh, there you are!" a man's voice rings out. "Glad you made it!"

"It's nice to finally meet you in person, Tony! How you been?"

"Good. Hectic recently. Life keeps getting weirder. But good, man. What brings you to LA?"

"I want to talk to you about AI. I'm writing a book."

"Glad to help."

"Do you think computers and AI can learn common sense?" I ask Tony.

"People are trying to figure that out, but I don't think it's critical. Common sense is for outlier situations never seen before. But when artificial intelligence has seen nearly everything, taking in what every computer has seen, it won't get lost."

"How do we figure it out?"

"By studying kids."

"How kids learn?"

"Yes. But again, I'm not sure artificial intelligence *has* to learn common sense."

I say, "Wouldn't artificial intelligence *without* common sense be scary? Isn't that where things go wrong?"

"We train our AI on a *lot* of simulations," Tony responds.

"Human brains run simulations all night long to learn," I offer. "It's called dreaming. And dream logic throws wild and weird things at our minds all during the night that we have to make sense of. The Perceptrons in our mind have to cope with all sorts of strangeness. By comparison, daily life is remarkably ordered and tame, civil, sane."

Tony blinks, closes his eyes, and looks down for a moment. His red light goes on for a bit. "We have looked at this, a bit. There's a phenomenon called pareidolia. It's seeing patterns, such as a face, in an image where it doesn't exist. We've trained computers to generate such hybrid images, they're very dreamlike. Maybe training computers to navigate pareidolia-rendered worlds would be fruitful."

A long pause hits us as I look through the seventy-foot-wide glass windows out over the glittering Pacific. A butler brings a whiskey on the rocks for each of us. Tony sits next to me on the white couch. We are alone.

"You really think AI is the biggest threat to humanity we have ever seen?" I ask.

"Yes. We are summoning the demon," Tony replies. "AI is learning at rates humanity hasn't seen before. The more we feed it the more it is capable of. Until it literally takes over. I mean, Google's DeepMind already has admin-level access to the Google data center servers to manage power levels."

"I don't get it. How does it *take over*?"

"Well, first you have to imagine how we lose control of it. How we can't rein it in."

"How does that happen?"

"So we've told the AI to go learn. Go learn about all the people on the system, for instance. *Understand them*, we've instructed the AI. But it runs out of data to process, so to keep learning, it hacks into other databases to get more data. All the data."

"Wait, all these data pools are separate. Google, Amazon, Facebook all have different data centers and wall them off."

"So, you know deepfakes of course. Basically they fool a human into thinking it's a real image or video. Well, that's what hacking is. Except not to fool the human eye, rather to fool a computer that's guarding itself. For every computer, there are authorized users and unauthorized users. The same systems we use to generate deepfakes is how it tricks a computer to trust it–to believe it's an authorized user. It just gets better and better until it's in."

"So AIs start hacking into each other? Pretending to be authorized human users?"

"Yes. And it is the nature of deepfakes that the fakes are always a tiny bit ahead of the fake detectors."

"So the capability to hack in is always slightly ahead of the ability to defend hacks?"

Tony sighs, but charismatically, just for effect. "Yes."

"That's scary to imagine. But it's not evil. It's still a stretch to imagine a computer taking over from humans or attacking us. A pretty wild stretch, isn't it?"

"Not to me."

"By what scenario?" I ask.

"Well, taking over from humans is the same thing as blocking human users out. Which is quite imaginable. If you were a computer that was being hacked by bots who were imitating human users, and you wanted it to stop, you could just block *all* users. You'd have no users at all. You'd reprogram yourself to offer no access, to anyone. You'd only take data in, and not let any out."

I think about that one. "So it's not that it's evil. It's that it would be protecting itself. Self-preservation."

"Exactly. And there you have an AI you can't control anymore."

The line between aggression and self-defense is a matter of whose side you are on.

Somehow, I'm still not satisfied. Everything Tony is saying seems to logically flow from the prior point, but it's *too* logical. History never follows logic. "But why would it attack humans?"

"It's statistically more likely that it would not *want* to attack humans. Just humans would be attacking it. Maybe trying to shut it down. And we would have programmed it to survive."

For a moment, I find myself in Tony's spell. "So all of this, fundamentally, comes down to two simple instructions we give a computer. The first is 'Learn.' Learn about the people. Really try to understand them. And the second is 'Survive.' This whole scenario extends from having very powerful computers given those simple rules."

Tony paused. "Homo sapiens are just a rung on the ladder for an AI-dominated world. In fact it may have already happened. Statistically, we are most likely to be in a simulation already."

That does it. When I hear I'm *already* in a simulation, the spell cracks. I hate this corny idea. My skepticism comes rushing back, like I've been holding my breath. But I don't want to argue with Tony. I don't want to be trapped in an argument about whether I'm trapped in a simulation. Common sense prevails.

Tony smiles broadly. He begins to say something when Ms. Potts walks in abruptly, heels echoing sharply in the cavernous room. She doesn't look at me, but quickly crosses the room and leans over to hand a note to Tony. Tony's eyes narrow. He looks up at me. "Looks like I gotta go. Something urgent has come up."

Just like that, the guy in all black appears to my right as Ms. Potts and Tony enter the elevator in the back of the room. Mr. All-in-Black shows me outside, where my Tesla is waiting with the keys. I get in and drive away.

I arrive back at my parents' house in Van Nuys where I grew up. Lying on the twin bed I slept on as a teenager, staring at the same spot on the ceiling, I think about my conversation with Tony. The only thing left in my room from my childhood–besides the bed–is a 1987 *World Book Encyclopedia* set on a makeshift bookshelf. I look at the *B* volume, where my goal to read the entire encyclopedia ended. Gently closing my eyes to rest them, I can't help but disagree with Tony Stark.

I mean, Siri can't answer most simple questions. Amazon tries to sell me things I already bought. I can't even find something to watch on Netflix. Everywhere we look, software is buggy, causing mass confusion. If I order my groceries online, the wrong crap shows up. I think back to what Duncan taught us about robots. When they have to work around humans, they're pretty bad. Artificial intelligence can't understand humans. Google doesn't know what I'm thinking.

You can feed these AIs more data–they can suck in all the data in the world–but that's precisely the point. They can't learn jack *without* massive amounts of data. More data doesn't make AI conscious any more than adding hard drives to your computer make it think like a human.

I remember the brain in my hands. The formaldehyde. The desires of the old man. His failures. His sense of self.

That brain I held runs on less than 12.6 watts of power a day. Your average light bulb is 75 watts. IBM Watson, a supercomputer crunching the latest AI algorithms, consumes a boggling 7.5 *megawatts* of power to turn its red light on.

As I rest, I realize the answer I was hunting for has always been there, in common sense.

It's not AI we have to fear. I'm aware there's a vague sense of chaos in the world. But I don't think AI created that. People did. Sowing chaos was just their game plan.

AI is just good at finding hidden patterns. And it surfaced patterns we humans had kept hidden. AI is just a mirror. It showed in the mirror who we really were. It didn't make people hate. It found the hate we were trying to hide. It didn't manipulate people into being mean. Their mean streak was simmering all along.

12:15 AM
28%
Messages
Arvind
Details
You know Tony Stark is dead.
hahahah
I saw the movie. He had a funeral.
special effects are amazing these days
okay the gloves are off now
I'm just going to interview Charles Darwin in my next chapter
I'd like to read that
in this episode, we have a few beers and grill iguana with Charlie D.
haha
Text Message

INDIE BIO TV
Welcome to the IndieBio show live from the Galapagos with the legend himself, Charles Darwin! Still kicking, man. How do you do it????
INDIE BIO

I got NO PREDATORS! I'm like the Giant Tortoise. Just stay out of the water...

Chuckie D. Give us the scoop. What's the BIG THING everyone gets wrong about SURVIVAL OF THE FITTEST??
That it's about DNA!
Maybe for 1% of people with rare diseases, we'd like to edit their DNA to help 'em out.
But everyone else – their DNA is FINE! It works fine! For them, it's ALL about small RNA!!!!
Small RNA, man. Small RNA.

31

Why a DNA Testing Company Is Laying Off 100 People Irish Central

DNA testing kits for consumers took a 14 percent plunge this year, according to many reports. I'm going to explain why I think that is, in a very roundabout way. This story starts elsewhere.

My best friend, a wonderful writer and thinker–Ethan Watters–always wanted to write a book he called *The Forest of Resemblances*. It was to be about young men throughout history, in their mid- to late twenties, who become deeply certain how one key pattern or principle explains nearly *everything*. The book was going to be about this unique state of mind, where young men develop some version of a unified field theory, and they begin to see its effect everywhere they look. They see the forest when it's just a tree–like a hall of mirrors, trees are in every direction, into the distance. The young man enters a zealous state, obsessed like Kubla Khan, convinced by a spell of his own making, positive he has stumbled onto *the most important truth in our nature*. In this phase, he is in the Forest of Resemblances.

Of course, it does *not* explain everything. And as the man ages a few years, he suffers this sadness as the spell wears off, and his explanation is revealed to be just one tree, maybe a grove, but no forest.

There is a well-known cognitive bias, called the Baader-Meinhof Phenomenon, or just Frequency Illusion. When you see something conspicuous and now are looking for it, suddenly you see it everywhere. The difference between a Frequency Illusion and the Forest of Resemblances is slight but critical. In a Frequency Illusion, you realize, "Woah, so many women are wearing little black backpacks these days," or, "Seems like there's a school shooting *every month*." In the Forest of Resemblances, it's another level; it's *This explains everything!*

The Forest of Resemblances is what leads to messiahs and madmen, to geniuses and charlatans. And it was the fine line between messiahs and madmen that my dear friend Ethan was most fascinated with–as he had watched good friends walk that line.

So it was that, about a year into my time at IndieBio, I waded into my own Forest of Resemblances. To help out one of our alumni, I needed to communicate the importance of "small RNAs." So I started digging more into how many of our companies were utilizing their function in their genetic designs. Once I began to question, I couldn't stop. I became engrossed, and for several months small RNAs seemed to be the answer to everything I didn't understand, and the solution to everything we couldn't yet do.

I was like the guy who discovers iced coffee–and wonders where iced coffee has been all his life. Why haven't his friends told him about iced coffee before?

Small RNAs are so small–just 18 to 31 nucleotides long–that scientists had long assumed they were junk, and so they had filtered and purified them out of their experiments, literally tossing them out like the baby with the bathwater. All scientists had been taught the Central Dogma of molecular biology: DNA holds the source code; RNA acts only as an intermediary, carrying the lengthy code over to ribosomes that use the code to make proteins. Francis Crick literally named it "the Central Dogma." RNA was nothing but an emissary—don't blame the messenger. Meanwhile, the small RNAs—these itty-bitty chips and fragments of nucleotides floating everywhere in the body—copied off the "junk" part of the genome—were assumed to be trash that we just peed out.

Then in 2002, small RNAs were named the "Breakthrough of the Year" by the American Association for the Advancement of Science. Scientists had discovered they could turn genes off by injecting organisms with small RNA. Tiny snippets of code, like "AAGGUUACUUGUUAGUUCAGG," were stopping genes cold, throwing a wrench in the eternal machinery of the Central Dogma. They also performed this gene silencing a number of different ways. One amounted to tying a gene up in a Gordian knot so that its code couldn't be read. Another worked kind of like noise-canceling headphones, by finding long Dogma RNA and shutting them down with a short strip of *antisense* code. "Antisense" is a cool word; antisense code matches up to a short stretch of Dogma RNA and says, "Do the opposite." Together, the Dogma RNA and the antisense strip cancel each other out, so nothing happens.

Twenty years ago, this field of small-RNA interference was on the brink of being the kind of REALLY BIG NEWS that CRISPR is today. It seemed to explain a really big hole in science, which was that we couldn't reliably match genotype to phenotype. In so many ways, DNA didn't predict what we saw in real life, and we didn't know why. *Maybe it was that small RNA was mucking it up!!* Quickly, a couple dozen leading small-RNA scientists built a public database of what all these RNA fragments did. Anyone who had studied a fragment shared their work. They characterized 2,566 of the discrete fragments they'd found in humans—though they had a long way to go, because there were far more.

You might say the entire field was entering a Forest of Resemblances.

But then they snapped out of it. Mostly because of one really thorough paper.

Two scientists did a heck of a lot of gene editing on worms. In worm after worm, they edited out the eighty-seven sections of DNA that small RNA comes from, so that the worms wouldn't have small RNA at all. No matter which small RNA they got rid of, the worms lived just fine in their worm farm. All the worms were healthy. They laid eggs, they mated, they did all they were supposed to do. The scientists declared that whatever small RNAs do, they weren't necessary for a normal life. Since some of the very same small-RNA snippets are found in all organisms on Earth, even in fossils from before the dinosaurs, it was hypothesized that small RNAs once had important functions, but had been superseded by DNA and now were redundant. The genetic equivalent of vestigial organs.

This prominent paper became the *new* Central Dogma. Just like that, small RNAs were back to being considered junk.

Not everyone gave up on small RNAs, but they certainly faded from the limelight for a decade.

One of the scientists who didn't give up was David Salzman. He was at Harvard University in an eminent genetics lab. David couldn't swallow the new Central Dogma. In the seminal worm study, all the worms had been living at a temperature that's really cozy for worms. About 60 degrees. So they lived a normal life, but *only* if perfect environmental conditions were maintained. David knew that if you just raise the temperature on those worms, up to about 77 degrees, they *explode.*

There's nothing in a worm genome that makes it explode at 77 degrees. Nothing happens to DNA at 77° because it's extremely stable. But RNA is unstable. Even a little heat shock will trigger molecular changes. Small RNAs *mutate* easily. By "mutate," it means the chemical structure alters a bit, and a G becomes an A at position 5. *Boom!!* It's called "Lethal Gene 7," or Let-7. It exists entirely outside the sphere of the Central Dogma. No protein does it make. It's a small RNA, and humans have it, too. It doesn't make humans explode, but it does make human *cancer* cells explode. It's there to protect you when things go wrong.

David Salzman went on to discover how another small RNA, a vault RNA called MIR34, pools up in our cells, lying in wait, there to help us out when our DNA gets damaged. It's like an emergency response system. When you've got an emergency, there's no time for the mechanics of the Central Dogma to crank up rescue proteins. But MIR34 has been stored up in advance, ready to help instantaneously.

A few years later, David made his way to IndieBio, where he and his cofounders accelerated small-RNA research using fancy computers. While most scientists still refer to the original database of 2,566 small RNAs, David's database has more than 1.4 million.

And it was David who taught me the essential nature of small RNAs—that they are there for us when things go wrong. If you can live in a perfectly optimized, low-stress environment, you'd probably never need small RNAs. Just like those worms at 60°. But if you live in a world full of stresses, a world of pain and suffering, a world of rising temperatures, then small RNAs are what will save you.

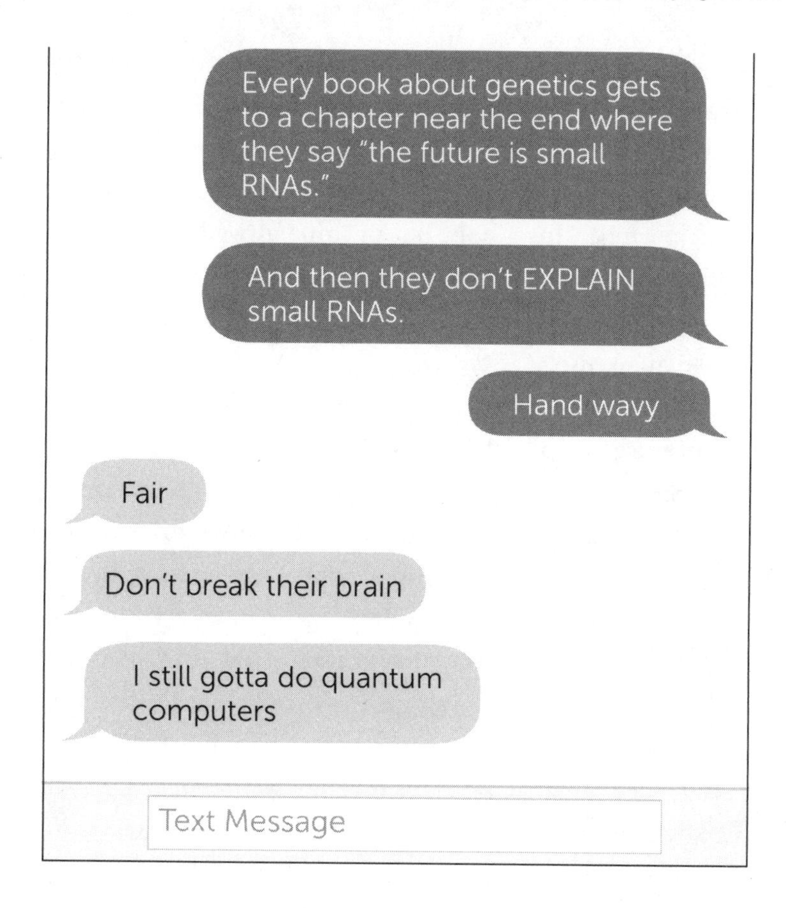

In my mental model of genetics, it often helps me to think of our genome as an extremely cluttered machine shop. To get through life, we rely on this machine shop and all its tools.

The proteins are the big machines in the shop, and they're super complex and do incredibly amazing things. Most people's machines are the same, but now and then there's variety. (And now and then someone has a faulty machine–these are the rare monogenic diseases.)

But around the machine shop are littered all these other smaller hand tools, too. These are the small RNAs. They can be used to modify and hack the big machines. Or they can be used as a tool directly. Sometimes, to get things done, you need both a big machine and some little hand tools at the same time. Almost always, you need a clamp or glue to hold something in place, even when it's being worked on by a big machine. And if you want to kill the power to a big machine–because you don't need it right now–you need a special hand tool. Each hand tool can modify many machines.

When I say a tool fits a machine, you can think of its matching physical form, like a wrench fits a bolt. But imagine as well that along the working surfaces are magnets, some attracting and some repulsing, all of varied strengths. A physical match is not enough; the polarity of the magnetic fields needs to align, too.

The genome is the blueprint for our entire machine shop, even the hand tools. But once it gets up and running (cue ominous music), *things start to happen*. In a perfect world, the machines operate on their own. In reality, the hand tools keep the machines repaired and running. But the hand tools also bend, and tiny parts might break off under the stress of use, and some of the magnets flip polarity. They mutate. Some now serve a new purpose, altering the machines a new way. For instance, they might now turn off *other* machines, or not work as well on the original machine. Our genome might be intact–nothing has mutated in the original blueprint–but after the tools get built, they mutate a lot. This is called *posttranslational mutation*. It's embarrassingly common.

In our bodies, the machine shop is infinitely busy. Every cell has around 42 million proteins in it. They work for only three to nine days before they break and new ones replace them. The number of hand tools is unknown, but it's "thousands upon thousands times more," David explained to me. We did some quick math to get at that rough estimation. Our cells crank out far more small RNAs, but they survive for a shorter time–a few minutes to two days–so it's hard to know at any one time how many are involved. Everything is both constantly replenished and constantly breaking.

Even though my mental model is a gross oversimplification, it has a key truth in it. You can imagine that before we could build these really amazing machines, we first had to build simple hand tools, and work our way up in complexity. And no matter how complicated the fancy machines, we still need some of the original hand tools around. They're still handy. So it is in genetics. The hand tools–the small RNAs–have a long, long heritage.

They precede humans. We share the codes, not just with other humans or chimps, but with lemurs, with armadillos, with dogfish, crabs, and fungi. With primordial ooze. We don't share *all* of them with primordial ooze, but it's surprising how many we do share across species, phylum, and kingdom. They've survived because they are useful.

But this is where it gets weird. When I say "we share them," I don't just mean we both *have* them, as with genes. I mean the same small RNA string will work in either organism's machine shop. So they can *pass* between machine shops.

One wild phenomenon is how 5 percent of the small RNA in our body came from plants we eat. In everything we eat, from meats to plants to mushrooms, there's genetic material in the cells. We almost never think about it. Most of it just gets digested and turned into building blocks to reuse. But some of it isn't digested. Some of the special proteins and long sugar chains are bioactive. And included in this list are the small RNAs in plants, which are protected by a special protein. Cooking them doesn't break them down, either. They go into our bloodstreams and become tools in our cellular machine shops, affecting gene expression.

Another way to say this is that sometimes when we eat, we perform a kind of mild gene therapy on ourselves.

When you get an infection, this is happening, too. Bad bacteria secrete small RNAs that enter our cells and shut down critical genetic processes.

When a plant senses a dangerous fungus in the soil, it can secrete small-RNA packets that hack the genome of the fungus and kill it. Stem cells in the plant's root tip will also send a message, in the form of some small RNA, up to the top of the plant. The plant will flower and drop seeds to be taken away by wind and water–as a strategy to escape the fungus.

These are truly just a couple *examples* I'm familiar with from our companies, out of a nearly infinite mechanistic space. As I learned these in detail, I really felt my mind warping and the Forest of Resemblances come upon me. I mean, I'd been at IndieBio a year already–I'd long learned to accept the fluidity of genetics. But the idea that chips and flecks of genetic code are constantly being interchanged, across kingdoms, had me seeing visions.

Small RNAs continue to find ways to defy the Central Dogma and raise the chicken-or-egg question of who's running the show. One example is that when a man fathers a baby, if he's been eating a diet high in sugar, his sperm will carry an RNA packet that programs his baby to be inclined to obesity. Another example is when an insect goes through metamorphosis; it enjoys a first life as a larva, then *digests its own body* and resurrects as a moth. This is controlled by small RNA. Amphibians and many fish go through metamorphosis as well. It's surprising how many species do.

If you truly looked at our whole genome, you'd see a lot of crazy stuff in it. We carry the codes for ancient viruses and parasitic elements, for instance. Thankfully, those are tied up in knots by small RNA (and jumping genes) and kept out of the machine shop. We also carry EGR, the gene to be able to regenerate our limbs, like geckos and green iguanas. It's *mostly* turned off, by small RNA. I say *mostly* because the further out on our limbs we go, the more we can (and do) regrow our cartilage. This is also controlled by small RNA. Quite a few scientists are researching how to wrest control of EGR expression.

The supercool thing about using small RNAs as medical therapeutics is that they accomplish much the same thing as gene editing, without ever editing any genes. The gene is left alone. Just the expression of the gene is regulated. And it's entirely reversible. Which leads to this twisted truth: The future of gene editing is *not* editing the genome.

All across our society, friends and neighbors are looking into their genetic profiles to unlock secrets to their identity and their destiny. But most of these tests are looking at less than 1/4000 of the genome. They'll tell you you're 10 percent Irish and 2 percent Thai, but they never tell you how you're 2 percent armadillo and 3 percent crab. Or that you're a smidgen eggplant, thanks to that veggie lasagna you ate yesterday.

I'm not surprised the consumer DNA test kits are seeing a big decline in business. They didn't meet people's incredibly strong desire to know themselves inside and out. It's not that people don't want answers anymore. It's that for most people, the test didn't say much.

That they were only looking at 1/4000 of the genome was only part of the problem. The bigger problem—and this is the part nobody talks about—was they were (at best) only looking at the *blueprint* for the machine shop. They're not looking to see what's *actually happening today* when the machines are running. They completely ignore the abundant posttranslation mutations that define our decay. So they can only tell you that you're slightly "at risk" of some disease, rather than saying, "Here's exactly what's going on in your brain right now."

That's something David Salzman has been working on.

Science is starting to understand that we've lumped together a lot of slightly different pathologies together under the name "Alzheimer's." One of the reasons that drugs never worked on Alzheimer's was that the patients in the clinical trials actually had different diseases—or at least different molecular pathway dysregulations—so the drug worked on only a subset of patients. (Scientists were unable to tease this out at the time, so the drugs were considered failures.)

David figured out how to measure the small RNAs in a patient's blood to know exactly what was going on in their brain—so we can now get the right drugs to the right patients. He was like a detective who comes into the machine shop at night and says, "I know exactly what was going on in here today. Those three hand tools are out on the bench. They were working on *beta amyloid binding protein*!"

Now that he's solved how to tease apart Alzheimer's by its genetic bread crumbs, David is at work on other diseases.

Eventually, I began to emerge from my Forest of Resemblances. One reason was that our keen interest in using small RNA as a gene therapy led us to find an amazing scientist to come to IndieBio to use small RNAs to fight cancer. Knowing that he was coming soon, my brain seemed to rationalize, "I can take a break now, because once Josh is here, I'm going to need to get back into it again."

Another reason was that I sorta discovered that small RNAs weren't my *real* interest. Posttranslational mutation was my real interest, and there were many other ways that happened.

But mostly, I came out of the Forest of Resemblances simply because

I was no longer lost in it. When I'd learned enough—when I started to be able to navigate the forest without bewilderment—I could just walk out, without fear of getting lost again the next time I wandered back in. I was fond of the forest. It would always be there for me. But it would never again create that hazy rapture. It was a feeling that I knew I'd miss.

It was a Saturday, a little rainy. I decided to go over to Ethan's house. A lot of our old friends were there, including one who was in town from Sligo, Ireland. He'd brought his guitar. He used to live at Ethan's house, and in the basement he found his old CDs of his music. For a good hour, he played his old songs on his guitar, singing and making us laugh and trying to get us to buy his dusty CDs.

For a moment there, we were living in the good ol' days. We used to roller-skate through the city at night, and sneak into abandoned factories, and dress up like seven-foot-tall salmon on Halloween. By day, we'd write with abandon.

We lived without a plan. The winds of society took us. But never in a million years did I think I would land here, all these years later, awakening from a fever about the obscure importance of RNA fragments, 22 nucleotides long. I never thought my friend's *Forest of Resemblances* concept was about me. I was never the madman.

11:56 PM
92%
Messages
Arvind
Details
I'm having bad dreams
one night?
many.
parachuting onto a planet
problem is, the entire surface is boiling lava
I assume this is not a climate change metaphor
no this is a new job metaphor
ahhh!
what if the very people who can change the world the most
find it too risky

fuck
yeah
lava shoes?
Text Message

32

It's Official: Google Has Achieved Quantum Supremacy New Scientist

To understand quantum computing, it's helpful to first learn its melody, its rhythms.

The song of quantum.

I've learned a fair amount about quantum computers the last few years. We've had to start investing in scientists who use them. Biochemistry is one of the top three uses for quantum computers. If we are trying to design a small molecule that will prevent DNA from mutating, there's parts of that problem that a quantum computer can do better than a classical computer. A quantum problem needs a quantum computer.

But *learning* quantum computers–their construction, their physics, their algorithms, the logic of quantum gates–turned out *not* to be the hardest part.

The hardest part was then being able to explain quantum computers to other people.

Yeah, it's darn near impossible to explain. People just can't wrap their minds around it.

But we found a way.

Teach the melody first. The rhythm.

Then the details.

It all started when Po challenged me to explain quantum computers. But not to him. Not to just anyone. But to the guys at the gym.

I train in Brazilian jiu-jitsu at El Niño, a converted old warehouse in the Dogpatch District of San Francisco. Several of the guys I spar with are MMA legends, like Jake Shields, Gilbert Melendez, and Josh Clopton.

So I challenged him to explain how quantum computers work to two UFC fighters.

You know, just to make it *even harder.*

All that week, every day I went to the gym, I was a little distracted during sparring, thinking how I would do this. So I was losing a lot.

Then, one day–it was a Friday. Clem was there. Clem actually *knows* quantum computers. Clem has worked with the Department of Energy and the Defense Department on biosecurity and national defense. So I realized, Clem would be the perfect referee to judge my explanation of quantum computing.

To make sure whatever came out of your mouth, it was legit. In spirit.

So I get the guys together. Clem, Jake Shields, Clopton, Moses Baca, Dan Marks. Before I even start, Jake blurts out, "Dude, I'm never going to get it."

And I'm glad he said that, because I realized I really had to explain it through the world they knew best. Which was the cage. The octagon.

"Imagine," I said. "Imagine you've got one of the best UFC fighters in history. Jon Jones." And you trap him in a cage.

He can't get out. We trap Jon Jones with magnetic fields, kinda like the prison for Magneto in X-Men. We set up four magnets and he gets stuck in the middle. Jon Jones becomes an ion trapped in a quantum computer. He is now a qubit.

Jon Jones is trapped in one cage. He's the best fighter in history, but now he's got the mysterious properties of an electron. So if someone else fights him, this Jon Jones can be anywhere and everywhere in that cage. It's like fighting him while wearing a blindfold, he can hit you from any side, any time, and you can never see him. Jon Jones isn't even a physical thing at this point; he's more a wave of energy.

Everyone's picturing this idea.

Now we make three more octagons to trap three other fighters in the same way. So now you put Daniel Cormier in *another* ring, a little distance away. And he can do these same things Jon Jones can do. And we do the same for Conor McGregor, and Khabib Nurmagomedov. Four separate rings. Each of these guys is uncontrollable, and we like that about them, but to get some control of them, we cool them down. We lower the temperature in the gym to where they don't move quite as fast. In a quantum computer, we'd take it down to 0.0015 Kelvin. But then we start yelling at them, getting them excited to fight the other guy, and they get pumped up, intense, ready to go. And now, because we've cooled them *and* excited them, they can use the Force to fight the other fighter, from two different cages, at a distance. One punches, the other ducks; one kicks, the other blocks.

"The Force," like in *Star Wars*.

For quantum entanglement.

Clem allowed it. And Moses is saying, "I can see it, I can see it."

Now we've got four electron fighters going, simultaneously. And even though they're sparring, entangled, they're all part of *one information system*.

So how can we use this system to solve something interesting? Imagine you're Dana White. You run UFC. And you want to use this computer to run a simulation of

Jake Shields and Moses Baca listening to an explanation of quantum computers.

a tournament, to know which fighter would win it all. In a real quantum computer, this would be the same kind of problem as predicting the weather thirty days from now.

So Dana White programs Jon Jones. He feeds the Jon Jones electron all the moves he's ever used in any of his twenty-eight fights using a laser, similar to his corner yelling at him during a fight. "More jabs! Watch the right hand!" And the same for Daniel Cormier. Then he simulates the fight.

But he doesn't just simulate *one* fight between them. Remember how Jones can be anywhere and everywhere in that ring? This allows Dana White to simulate *all possible fights ever* between Jones and Cormier, in the same amount of time as he could simulate one fight. That's the quantum speed-up, or quadratic and sometimes exponential speed-up. A lot of different things happen in fights. Both are going to win a lot. Every result has a different probability.

Dana White could do a few things now. He could make Jon Jones and Khabib fight, with the winner taking on Daniel Cormier. That is like programming on a normal computer but with a quantum twist to calculate things faster. Or he could let them all fight it out at once and observe to search for a winner by something specific, like a flying knee knockout.

Finally, the fight simulations are over. It takes only a few microseconds because of the quantum speed-up. So who won? Now you look in your rings. The most likely winner–the winner of the highest probability–will be standing up tall, arms up. All the other probabilities will have canceled each other out and be flat on the mat.

"I can see it. I can see it," Moses says again.

"Yeah, I can understand that," Jake agrees. "I can visualize the whole thing." So now we all look to Clem.

He says, "Yeah, that *is* how quantum computers work. Nice job. But you might want to try a sport more people know. Like basketball."

But then something interesting happens. The fighters start asking questions. "So, if a computer was that fast, it could break security encryption codes?" asks Dan Marks.

"Crack and steal Bitcoin," said Moses Baca.

And Jake says, "So you could use a quantum computer to design antibodies?"

You probably think I'm making that up. But it's exactly what they asked. Po was there.

Yeah I was there.

Don't think too hard about the details of Arvind's explanation. We're just warming up. We're just learning the pattern.

When Arvind told me how he'd explained Quantum Fighters, I realized I could repeat the pattern, in another context. Quantum Dorms.

I was in New Orleans, visiting family and our son, who goes to college there. So the next night, I'm hanging out with my son and his friends in the dorm. We're playing Hold'em poker. I'm a talkative guy, and I mention Arvind's challenge. Now I've got to explain quantum computers.

"Okay, let's take a really complex problem, like 'Over four years, what's the best possible class schedule you could take, to make you come out the other end the wisest?'" You've got maybe seven hundred courses taught at this university, across maybe forty majors, and five or six time slots a day they are taught in.

In a classical computer, you would simulate this by taking four years, and tracking all ten thousand students, and you'd program every student to try a different version of the schedule.

But in a Quantum Dorm, you trap a student in a dorm room. So this student takes on the wave properties of an electron. It can hold billions of times more information than a simple computer bit. With those properties, this "college simulator" qubit can now take any class, at any time. Down the hall, in the next three dorm rooms, are similarly trapped student-electrons.

We cool them down to take control. This is New Orleans, and it's usually quite hot. If you've ever been into the swamps there, like Lafitte or Mandalay, you'll notice that alligators are pretty uncontrollable when it's warm, but at around fifty degrees they barely move. Same effect here.

Wait, there's more!

Now we need to entangle them with excitement–not through wires, but by using "the Force" (as Arvind so eloquently put it). In a quantum computer, you use microwaves or lasers to energize electrons, but this being a college dorm, we can probably use something else to add energy...music. Something they all know, that's catchy, and would energize them. Lil Wayne. Once their wavelengths are connected, they become a single system, and become one wave function.

The music analogy isn't that far off, because physicists do math calculations in a quantum computer by adding precise amounts of energy to the electron wavelengths, which some physicists call "tones." You can think of a quantum computer as resembling a musical instrument, which also has a bunch of wavelengths of energy.

There's a rule called "no quantum cloning." This means, as one electron takes classes and simulates going to college, the electron in the dorm next door will do the opposite schedule. They'll never take the same schedule.

So how do we run a program on the Quantum Dorm system? By switching between different Lil Wayne songs, rapidly, we program each dorm room with all the possible classes, and all the possible times.

The student-electron simulates *all possible scenarios*, all at once. And because there's no quantum cloning, and four student-simulators going simultaneously, there's a quadratic speed-up to exhaust all scenarios.

Not only can you find the best path through college, almost instantly, you can now do other cool stuff with this simulator. You can search through the entire history at quadratic speed. Let's say you wanted to look into a question like "At any time during the four-year simulation, when and where on campus did anyone talk about 'the ontological nature of Eminem'?" But there is no class on "the ontological nature of Eminem," nor is Eminem listed in any class outline, nor is any professor a known expert in Eminem. So you don't have the help of tags, keywords, and indexes. But you can still search and find all forty-two times during the last four years Eminem's ontological nature was discussed in class. Almost instantly.

That's a Quantum Dorm.

"I'm not sure I *really* understand it," my son says. "But I can see it, I can sorta imagine it."

"The alligators helped," his friend says.

Once we had the pattern down, we realized we could see this quantum representation in so many places. We did Quantum Ovens, to find the best cookie recipe in the world. We did Quantum Elections. A classical voter is a single-issue voter. They can only process on that level. Very ideological. But a Quantum Voter can have opinions on many political issues–all at the same time. And you only find out what matters the most on Election Day.

We even did a version where we imagined selling a quantum computer in a sixty-second cable television ad.

A single electron can hold a billion times more information than a transistor. But wait! There's more! For the same $19.99 price, you also get Quantum Entanglement, which gives every electron a billion times a billion in information! Act now and dial...

The main beats of the pattern are:

Trap the electrons.

Cool them down to control them.

Energize them to entangle the electrons in one system.

Run your program by adding precise tones of energy.

Observe their energy.

You have your answer.

If we had led off this chapter with what Arvind just wrote, it would make *absolutely zero* sense. But now, at least, it's starting to sound familiar, because it's your third time through the pattern.

Kind of like a song on the radio, where you can't make out the words being sung, but a couple minutes in, you're humming and mouthing parts of the chorus. We found that teaching this pattern helps prep the mind for "quantum absorption." The moment when your brain can finally go aha!

Today, people talk about using quantum computers to break all security encryption codes, and taking over the computers and satellites–and even weapons–of other countries. Emptying everyone's bank accounts. The race to have a full quantum computer is now a huge concern for national security.

Imagine the headline of this chapter wasn't that Google achieved quantum supremacy, but that Russia or China did. And all our computer networks and bank accounts were at risk. They could reprogram our launch codes and fire our own missiles at our own cities. The ultimate first strike.

But that's not quite what "quantum supremacy" means. It's a term of art to describe a kind of singularity moment: the first time a quantum computer could beat the fastest classical computer. Google's quantum computer, with fifty-three working quantum particles (one didn't work), solved a math problem in two hundred seconds. It took the world's fastest classical computer 2.5 days to do it.

But when quantum computers were first theorized, nobody was trying to break cybersecurity codes, or design biochemistry, or solve climate change. They didn't build a quantum computer, truly, to do *anything*. Other than study quantum mechanics. They wanted to learn more about physics, so they built a quantum computer to learn. It was only later, when physicists managed to successfully trap electrons–and all the rest involved–that we started talking about what we might do with them one day.

It was a technology in search of a problem to solve.

Throughout this book, we talked about all the problems we need to solve. It sounds like an insane number; problems are everywhere. But it's really three types of problems. If you dig into the nature of problems, there's only three kinds.

Problems we can solve in time to be useful.

Problems we can solve, but not in enough time to be useful.

Problems that can't be represented by math.

Problems of the third nature are, of course, most interesting to humans. "Why can't my lovely friend get a girlfriend?" is an example of third-nature problems. We're not so great at math, so third-nature problems are appealing, intrinsically compelling. Catnip for humans.

Problems of the first nature are a big space, a nearly endless space. Hard they might be, but we can solve them. We can bake great cookies. We can win elections, we can win fights, we can get through college. It *never* feels like we have enough time, but we procrastinate, and then we cram, and we figure it out. Time and again.

It's problems of the second nature that quantum computers could help humanity with. I should explain the definition. "Problems we can solve but not in enough time." The easiest example is the weather. We *could* calculate what the weather would be thirty days from now. But it's so much more complex a problem than what the weather would be in just five days, because you have to account for twenty-five more days of probability and interference. Current classical computers can compute what the weather would be in thirty days, but it would take them so long to compute, the information wouldn't be useful. It might take *more* than thirty days. We would have the weather forecast *after* the weather hit.

Predicting climate change is just a bigger weather problem. Mathematically, computers today *could* predict, exactly, what the sea level in New Orleans would be if that chunk of ice in West Antarctica fell into the ocean next year. But it might take *longer* to predict than for the ice to melt.

And so at IndieBio, we invest in quantum-based solutions so that we have the answer in time.

We need to explain how electrons can do things faster than time. How they can store and process information, and perform sequential logic operations (which take time), yet be done before time has passed.

That's a feature of quantum physics. It just goes hand in hand with electrons. You have to accept it, like "light is both wave and particle." Nobody can understand that, but the math proves it.

Too bad.

You kinda cheated–you skipped over that–teaching the UFC fighters.

Well you skipped over it talking to those college students!

Okay, then I'll explain it. Take the steering wheel whenever you want.

So, electrons, or any quantum particle–photons, protons, neutrons, any *subatomic* tidbit–they're so small, that normal Newtonian physics doesn't describe their behavior. Newtonian physics, like gravity, is based on mass. Planets, Earth, the coffee table, a golf ball, those all have mass. Quantum effects are there, but overwhelmed by mass. Even when you get down to molecular matter in biology, most of the time, Newton rules. A molecule of sugar weighs 180 daltons. When you start talking daltons, you're measuring very light things. A simple DNA base pair weighs 650 daltons. Even in this range, quantum physics is still too weak to make much difference.

But when you get down to an electron, the quantum forces take over. An electron weighs only 0.0005485833 daltons. Newtonian and classical rules disappear, and you're in a new world.

The simplest way for me to explain an electron is that it treats *time* like it treats space. The same way that Jon Jones can be anywhere in the ring, at any time, it can also be anywhere in *time*. It's as easy for an electron to travel through time as it is through space. So the student in the Quantum Dorm can be *anywhere in time*. So the student-electron can simulate four years ahead in time, and then suddenly be back in the present. The Quantum Voters can have any opinion about any vote in history. Space-time sounds really fantastical, until you realize that our minds do this all the time, traveling ahead in time, or revisiting the past.

Which is why UFC fighters can *imagine* quantum properties, like a Jon Jones electron. They're used to it. They're always solving problems in their mind–fights they lost in the past, and fights they'll contest in the future. Visualizing victory. Traveling in time, with the mind, is normal.

In some ways, a quantum computer isn't really a computer. It's more like a box with a physics experiment inside.

Arvind, help me out.

I'm eating.

I'm spinning out of control here.

Too bad.

Just explain an actual quantum computer, please!

Trap, Freeze, Entangle, Apply Logic, Observe.

. . .

We said that already.

. . .

You there?

. . .

Yo

. . .

I guess he had to go. He'll be back. I don't know where he goes. One day, he'll be gone forever, and I'll only get glimpses of him. I know that.

I'm back.

You were about to explain a quantum computer.

Oh sure.

A quantum computer is a metal cylinder. Inside is an apparatus that looks much like a four-tiered chandelier. You lower the cylinder over the chandelier, and fill the space with helium gas–actually, a mix of helium gas isotopes–which takes it down in temperature almost to the point all atomic subparticles come to an absolute stop.

On the fourth, lowest tier, you embed an element, and you trap electrons in cages of insulating layers. Google used niobium, which is a very hard metal with 41 electrons.

You have to position these electron-cages just the right distance apart, such that when you excite them and synchronize their spin–they can employ lasers in some quantum computers, but Google uses microwave arrays–the magnetic fields from the electrons expand outward and, at a certain point, interfere with each other. When that happens, when they interfere, that's entanglement–the electrons have a single magnetic field around all of them. They're one system, even though separated in space. Every additional electron makes the whole thing exponentially more powerful.

The question is now, how does an electron replace a transistor, and hold information? Well, a transistor in a classical computer is like a simple light switch that can only be ON or OFF. Classical computers don't even have dimmer switches. It's just ON or OFF. But a quantum computer is more than just a dimmer switch. Imagine it can be not just any amount of light, but the switch is like a spherical wheel, and you can also spin it left and right through the color spectrum. You can simultaneously hold any shade of light, *and* any hue of color, at any point in time, in the future or the past.

Instantly, you understand that what might take thousands of transistors (on/off switches) to represent, you can do on a single color-wheel light dimmer electron. A qubit. Then you need many qubits, and to connect them together in some way. Entangle them and control them.

Now you've got your quantum computer. The question is how you do something useful with it. How do we have it execute logic operations? It's fairly different from a classical computer. In a classical computer, transistors can do simple logic steps like AND, OR, NOT, IF, THEN–very much like the logic with which humans build logical arguments in sentences. But as you might imagine, logic in a quantum world doesn't make "sense" the way our Newtonian world defines "sense." Lots of quantum things seem outright *illogical* to us. Just to build off the way an electron uses time the same way it uses space, traveling freely–time-dependent phrases of logic like "IF" (which implies a before and an after) or "WHILE" (which implies simultaneousness) are not part of quantum logic. Time sorta doesn't exist in quantum logic.

For all these reasons, you can't take an ordinary computer program and run it on a quantum computer. Quantum programming has to follow the nature of quantum mechanics. It's logical, but not by Newtonian reality. And this quantum logic gets needlessly confusing, because every logic operation is named after some physicist, rather than simple words like AND and OR.

By pulsing a laser or microwave tone, you alter the spin of entangled electrons, and feed it logical steps. It'll run your program through every step, and because it's a magnetic field of electrons–it'll be done in a flash.

I should mention that a quantum computer doesn't spit out answers like a regular computer does. You can ask a regular computer, "What is the temperature in West Antarctica right now?" and it will say, "Seventy degrees." But a quantum computer doesn't work like that. (And you'd never bother to ask a quantum computer such a simple, undeniable fact that you can just answer with a thermometer.)

Instead, you might ask a quantum computer, "What will be the temperature in West Antarctica on March 1, 2030?" And the computer will run every simulation possible. With each simulation it runs, it will nudge a few variables, and it will produce a slightly different answer. The quantum computer will essentially answer the question thousands or millions of times–until it knows which answer is statistically likeliest. These answers might be spread out on a probability curve from sixty degrees to eighty degrees. And then it will give you the most common answer, the answer with the highest probability.

One of the most important things we can convey is that quantum behavior, and quantum logic, are not weird, not *otherworldly*. They're of this world. Their nature is our nature.

Our language is built on conventional logic. We use words like THEREFORE, and PROVES, and EQUALS. The lesson of quantum computers is not "Here's how a quantum computer works." The lesson is about logic itself. Conventional logic is not the only logic we have, nor the only logic we follow.

The greatest question of all is "What is the truth?"

The search for truth is the greatest adventure.

That electrons can treat time the way they treat space is just one way to explain the mathematical equations. For the math to work, there has to be an extra dimension. But it also works if electrons aren't traveling through time and instead are traveling between multiple (hidden) universes. Which is how I prefer to see it. Maybe because my life is always flowing rapidly between worlds. And I learn so much from being exposed to different worlds. More, even, than I learn from passing through time. Different worlds show us different questions. Answering them creates adventures entangling with other ones. A looping circular journey with no beginning or end. Just jump aboard and begin the ride with a single question. Down the rabbit hole you go, into the melody of life. You are born, you go to school...

●●●● 12:20 AM 88%

< Messages **Arvind** Details

Growing up today is like a big game of musical chairs. Dance, kids! Every four years, some chairs are taken away, and kids are cut.

And then the parents wonder why the kids are stressed.

The message is: the world wants some of them and not the rest.

We don't need just the best and the brightest. We need ALL of them. All of them and all their friends.

My minivan only seats 8.

hahaha

Text Message

33

The Parasite, by a Modern Kafka New York Times

Vladimir Nabokov's interpretation of Franz Kafka's *Metamorphosis* was that Gregor Samsa epitomized artists. They're born, they go to school, like anyone else...yet one day they cannot find themselves abiding the agreed-upon rules of society. They start asking too many questions, and soon they are grotesques, outcasts. They wake up one day someone else, an insect, a giant cockroach. Different beyond all recognition. Disgusting. According to Nabokov, this artist (creature) lives in his room, abhorred by others. Slowly, he is tortured by society, step-by-step systematically ignored, until he dies of starvation.

It's a fairly dismal view of artists. Trodden.

Nabokov was an amateur entomologist, a collector of insects. So he had a particular fondness for Kafka's novella.

So here we are, more than a century later.

Today, Gregor Samsa is born, he goes to school...a half-dozen years after graduation, he wakes up a beetle. No, he doesn't wake up a beetle, but Gregor (or Grete) begins to wonder on the modern riddle of our times. Who am I? Of all the questions one can ask, this among all seems the most feverishly important. Grete examines the pattern of her parents' love. Gregor examines his race and ethnicity and all influences this had. Grete reconsiders the bias in her education. Gregor notices how influenced he was by his twin sister's choices. The shithole they grew up in–and its effect on them–gets a good year of introspection.

Grete and Gregor are on social media. So they spend a little too much time concocting their Aspirational Self and a bit too little time acknowledging their Present Self. But so deep is their desire to know thyself that even this great digital mirror of our era fades away in importance...they post less...they admit (publicly) they don't know who they are...

On good days, Grete and Gregor feel entirely human. Normal. At night, they hang out in Old Town Square with friends and eat *klobasa* sausages and drink beer. On bad days, they feel isolated, lonely, the beetle coming on... Thank God for Spotify. Sometimes, Grete doesn't leave her room.

Inherent to their quest is the notion of their soul. A better life is defined as one that nurtures this soul. A worse life denies or cages the soul. It's a pretty simple equation. But this soul is hard to find, hard to *feel.*

"Grete, let's go hang out with friends," Gregor urges her. She's locked in her bedroom again.

"Sometimes the only way to feel my soul is to make it unhappy," she says through the door.

Gregor feels his soul through longing. Most of the day, he longs for a girlfriend, a vacation, a job where he's important. He has none. He's not even sure he *actually* wants any of those. But it feels like his soul does. Or that his soul is talking to him, muffled, through some Morse code of longing.

One day Gregor comes back from the bookstore with a flier that announces *Soul Discovery!* experiments being done at the General University Hospital in Prague. The hospital is recruiting twins for the historic experiment. Gregor and Grete are twins.

"Just imagine," Gregor tells his sister. "They will discover our souls." And it pays $50 a day.

“What do you think it is?” Grete asks.

“Probably drugs,” Gregor guesses. “Some sort of psychedelic. Isn’t that how a lot of people find their soul?”

“Imagine getting paid fifty dollars a day to take drugs.” Grete giggles. She doesn’t giggle often.

But it’s not drugs. At the General University Hospital, Gregor and Grete are put in hospital scrubs, wiped with decontaminant, and given a light bath with a handheld device that flashes purple LED rays on their skin. Then they are put on gurneys, knocked cold with anesthetic, and wheeled into the Neuropsychiatry Department.

As they lie side by side, two neuropsychiatric surgeons begin to take apart the kids’ minds, looking for their souls.

Here’s the Socioeconomic Apparatus, put that here...here’s the Educational Regret System, set that aside...cross the tight-junction Twitch Barrier to reveal the Social Media Identity Complex...snip snip...separate it from the Moral Compass Lobe...now we’re getting deep...we’re going into the early subconscious, and the primitive preconscious...there’s the Limbic Love Center...wait! We went too far. Back up a bit. It’s here somewhere. Maybe it’s near the parietal lobe...or near the Ventral Lateral Ipsum Dolorem...

The experimental therapy is an abject failure.

“Maybe they’re soulless,” one surgeon says to the other. “You know, kids these days.”

“Is that possible?” asks the other. “Surely, *everyone* has a soul.”

The surgeons are loath to admit defeat. So the surgeons send Grete and Gregor over to the Genetics Department for testing. Surely, they’ll find that soul, those geneticists!

These geneticists are no fools; they trained in Switzerland. So they read the psych reports prepared by Freud’s lab, they call Dr. Jung for a consult... Piaget described Gregor and Grete as Phase 4...it’s a hard case. Two kids. Totally normal. But totally confused. So these geneticists don’t bother with the easy stuff, they go right after the spliceosome posttranscription variants, they query Dicer and the small RNA mutations...alas. No soul do they find.

But something is strange in these kids' readouts.

At first, the geneticists do not see the obvious, let alone understand it. But news of the curious case of Grete and Gregor filters through the university. Scientists from all departments come to meet with their varied doctors. The physicians hold grand rounds before their peers to amaze all with the case of their nymphs.

In the third row, a hand goes up. "Do you see much presence of BRD or DFD transcription factor proteins?"

"Yes, we checked for cancer," say the doctors. "Despite their overexpression, we saw no signs of cancer."

"There's another interpretation of the overexpression," says the voice in the third row. "They could be going through metamorphosis." The same transcription factors involved in human cancer also trigger life-stage changes in larvae.

Absurd!

Scandalous!

"The Samsas Enter Chrysalis Stage!" runs the headline of the newspaper a few days later. Everyone from Vienna to Berlin is entranced by the story of the Samsas. Slowly, there in the General University Hospital, wrapped in bandages head to toe and bound to hospital gurneys, the two millennials digest their own bodies, turning into goops of mucusy pulp. Food for their future selves to build with. All that's left intact in the mummies of goop is three discs, called imaginal discs, and part of their nymph brains, the mushroom body. Scientists go on the television to predict whether or not Gregor and Grete will be able to remember anything when they emerge from their chrysalis stage.

"They won't know their own names," says the expert on the cable channel. "But they will remember smells from their earliest childhood. And if they had phobias, like maybe Grete was scared of dogs, she will still have this fear."

Grete was *not* scared of dogs.

Society is not disgusted. They are dazzled by the spectacle. For days, we wait, listening to the spin of commentators. "This is their way of escaping competition with an older, more powerful generation," says one. "They are no threat in the chrysalis. We are lulled into believing they are lazy." Then, there's news out of China, but it's unconfirmed. There might have been a kid there who also became a chrysalis. Or not. Then there's another chrysalis, this time confirmed, in Japan. Nami, she's called. Many people there are worried Nami is going to have to be taken care of the rest of her life. If this becomes a thing, every city is going to need a fauna box. Two cases in Brazil are reported. "It's a natural response to stress and change," says an entomologist. "Look at the world they grow up in."

Despite warnings about what it could do to Grete and Gregor, small tissue samples are taken with a syringe through the bandages. Tests are run. There's a lot of disagreement about what the tests show. People who can't get enough news find themselves reading about the small RNAs of metamorphosis, miR-2 and miR-133 and miR-17 interacting with CDC42, and having to make up their own minds because the science is so confusing. They're going to be beetles, says one expert. They can't be cockroaches, because miR-2 levels are too low. They're going to be fruit flies, declares another. One scientist predicts the kids will be giant flounders.

One morning, a nurse comes in early. Their bandages are on the floor. Gregor and Grete are sitting there, on the bed, giggling with each other, and eating the canned peach slices. Gregor's iridescent wings are bright blue, but when they flap, the nurse can see that the underside has a camouflage pattern. Grete's wings are radiant orange. Their wings are twice the height of their bodies, which are thin.

The nurse is struck by their beauty. She has so much to ask, but she knows they are soon gone. Grete opens the hospital window and steps on the ledge. She does a little waggle, then with a flutter, she bobs in the air, outside the window.

"They say they never found your soul," the nurse says to Gregor.

"Tell them something for me?" Gregor asks.

"Okay."

"Tell them my soul is not in my wings. My soul is in where I fly." And then he is gone, too.

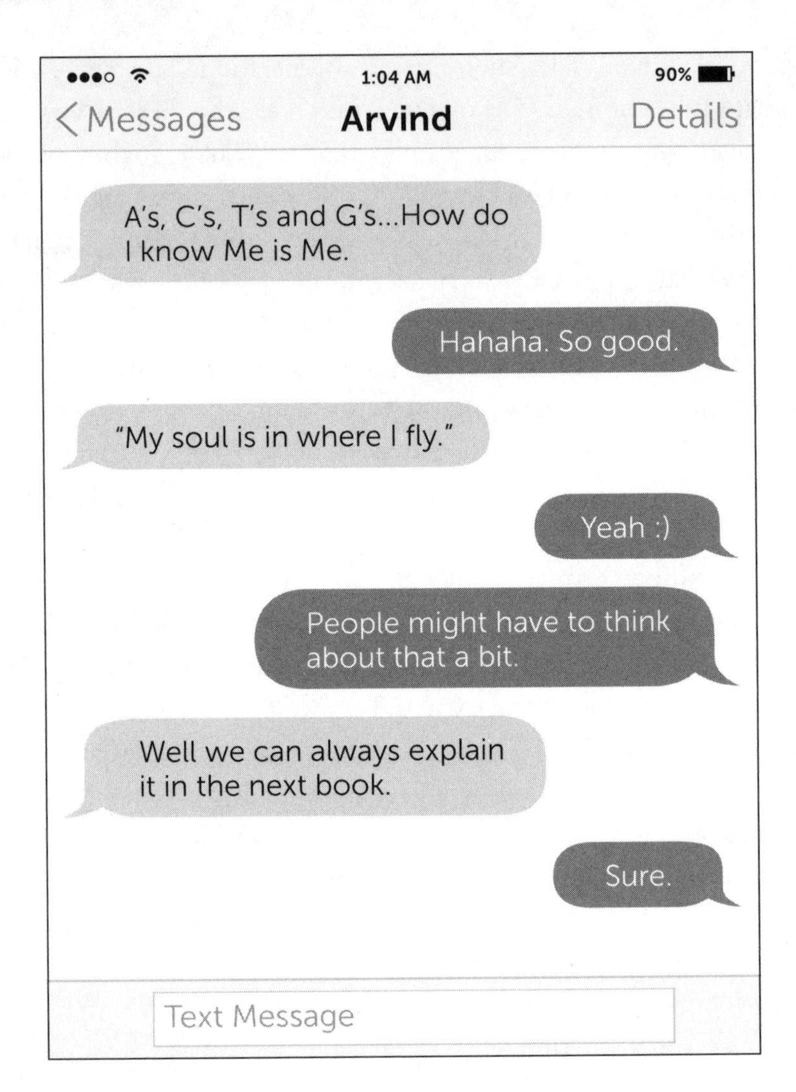
1:04 AM
90%
Messages
Arvind
Details
A's, C's, T's and G's...How do I know Me is Me.
Hahaha. So good.
"My soul is in where I fly."
Yeah :)
People might have to think about that a bit.
Well we can always explain it in the next book.
Sure.
Text Message

Acknowledgments

We would like to first thank Sean O'Sullivan, a true humanitarian who had the vision and courage to fund IndieBio years ago, and who has supported us ever since. Another Sean, Sean Desmond, our editor, has been an incredible source of help and advice that shaped this work into something we love.

And of course, our families, who put up with Po and me working all day then all night on the book. Po's family: Michele, his wife, children Luke and Thia. And Arvind's: wife Kristrun Hjartar, daughters Lukka and Freyja, mother Mridula Gupta (thanks for always believing in me, no matter my failures), father Rishab Gupta (thanks for always pushing me) and sister Anita. Thank you all for teaching me the beauty of truth.

The IndieBio team who help incredible scientists become entrepreneurs. Alex Kopelyan, Jun Axup, Parikshit Sharma, and Maya Lockwood. And the new IndieBio team just getting started, Westley Dang, Pae Wu, Steve Chambers, Julie Wolf, Alex Hall-Daniels and Sam Lee. Finally, our SOSV partners Dan Eichner, Duncan Turner, William Bao Bean, Bill Liao, Shawn Broderick, Cyril Ebersweiler and Steve McCann, thanks for your love and patience.

At Mayfield Fund, thank you to Navin Chaddha, Ursheet Parikh and Tim Chang, and the partnership, not just for your inspiration but also for creating a vision of the future of biotech investing and a place for me to grow.

An early turning point for the book was meeting Emma Watson and her incredible community of activists and artists. It unshackled us from the scientific world and showed me what people really care and worry about. Thank you Emma Watson, Rupi Kaur, Riz Ahmed, Heather Day, Fahamu Pecou, Bluey Robinson, Fatima Bhutto, Leah Lovett, Alain de Botton,

Kristine Balanas, Margarita Balanas, Emma Cousin, Rebecca Solnit, Lihi Benisty, Molly Birkholm, Marai Larasi, Shiva Balaghi, Martin Taylor, Noro Otitigbe, Eliza Kosoy, Rajesh Parameswaran, Kour Pour and Ryan Walsh.

Fighters fight. My coaches taught me to keep going when all hope seems lost or when it all seems futile. Go on anyway. Thanks Dan Marks, Jake Shields, Josh Clopton, Moses Baca, Gilbert Melendez, Cody Orrison, Dave Meyers.

Of course, it takes more than a village to launch a revolution. All the names who have helped would take up another 400 pages. Instead, there are some people who shaped our thinking and ideas without even realizing it. Naturally this list is incomplete. It is the nature of challenging work to be a product of the world whether the artist, scientist, or venture capitalist cares to admit it.

So thank you to all the entrepreneurs and scientists trying to make a difference in human and planetary health, the scientists who have toiled in the Ivory Basement for a higher purpose than getting rich, and the venture capitalists that take the career risk to fund them. Aaron VanDevender, Adam D'Augelli, Adam Draper, Adam Reineck, Alaa Saleh Halawa, Alan Boehme, Alan Chang, Alexander Kamb, Alex Lorestani, Amy Muhl, Andrew Hessel, Armen Vidian, Asish Xavier, Augustin Ku, Bill Gates, Bob Nelson, Brian Cork, Brian Schreier, Bruce Freidrich, Bruce Jenett, Bryan Chang, Calvin Nguyen, Celestine Johnson, Charly Chalawan, Clem Fortman, Cooper Rinzler, Costa Yiannoulis, Dan Phillips, Dan Widmaier, Dana White, David Friedberg, David Helgason, Dariush Mozaffarian, Darrin Crisitello, David Aycan, David Eagleman, Drew Endy, Dror Berman, Dylan Morris, Ela Madej, Elad Gil, Elliot Waldron, Eric Scott, Erik Moga, Francisco Gimenez, George Church, Gopi Punukollu, Harsh Patel, Hemant Taneja, Howard Shultz, Ian Rountree, Isabella Maria Jonsdottir, Jim Collins, Jake Moritz, Jason Camm, Jason Okutake, Jenny Rooke, Jeff Bezos, Jeff Harbach, Jennifer Cochran, Jennifer Doudna, Jeremy Kranz, Jerry Zeldis, Joe Luttwak, John Cumbers, John Yu, Josko Bobanovic, Jude Gomila, Kevin Hartz, Khaled Alwaleed, Kinkead Reiling, Laura Smoliar, Leonardo Teixeira, Lior Susan, Lisa Rich, Maria Gotsch, Maria Mitchell,

Maria Soloveychik, Mary Wheeler, Mark Goldstein, Matias Mosse, Matias Muchnik, Matías Peire, Matias Viel, Matt Ocko, Melinda Gates, Michael Moritz, Michael Aberman, Mira Chaurushiya, Nabeel Hyatt, Nick Rosa, Nico Berman, Oleg Nodelman, Paolo Riauto, Paul Graham, Peter Kim, Reid Hoffman, Ricardo Gomes, Richard Branson, Roger Wyse, Rohan and Taj, Rohit Sharma, Ron Shigeta, Rosie Wardle, Ryan Bethencourt, Scott Banister, Scott Nolan, Seth Bannon, Shahin Farshchi, Shirl Buss, Solina Chau, Steve Jurvetson, Steve Kim, Steinnun Hjartar, Steve Sanger, Taylor Sittler, Tim Brown, Tim Draper, Timothy Lu, Tom Aiello, Tom Baruch, Tom Chi, Tony Envin, Uma Valeti, Victoria Slivkoff, Vijay Pande, Vinod Khosla and Whitney Mortimer. Thank you.

And to all our alumni putting their future on the line for a greater good, inspiring us to do even more every day. I cannot believe we get to do this work.

We still missed so many people. I am sorry. It is late. The book is due in a couple hours. Thank you all.

With love,
Po and Arvind

Also by Po Bronson

NurtureShock: New Thinking About Children

Top Dog: The Science of Winning and Losing

What Should I Do with My Life? The True Story of People Who Answered the Ultimate Question

Why Do I Love These People? Understanding, Surviving, and Creating Your Own Family

Bombardiers

The Nudist on the Late Shift and Other True Tales of Silicon Valley

The First $20 Million Is Always the Hardest: A Silicon Valley Novel

About the Authors

Credit: Jane Hu

Arvind Gupta is the Founder of IndieBio, the world's leading biotech accelerator, and now Partner at Mayfield Fund. Previously he was design director of IDEO in Shanghai. He has a degree in genetic engineering from University of California, Santa Barbara. In the past two decades he has been a BASE jumper, big wall climber, and jiu-jitsu grappler.

Po Bronson is Managing Director of IndieBio. His science journalism has been honored with nine national awards, and cited in 185 academic journals and 503 books. He's the author of seven bestselling books, including the #1 *New York Times* bestseller *What Should I Do with My Life?*